IDIOT'S
GUIDES.
AS EASY AS IT GETS!

Geometry

WITHDRAWN

by Sonal Bhatt and Rebecca Dayton

ALPHA
A member of Penguin Group (USA) Inc.

To Sonal's mom, Gita, and Rebecca's mom, Bernadette.

ALPHA BOOKS

Published by Penguin Group (USA) Inc.

Penguin Group (USA) Inc., 375 Hudson Street, New York, New York 10014, USA • Penguin Group (Canada), 90 Eglinton Avenue East, Suite 700, Toronto, Ontario M4P 2Y3, Canada (a division of Pearson Penguin Canada Inc.) • Penguin Books Ltd., 80 Strand, London WC2R 0RL, England • Penguin Ireland, 25 St. Stephen's Green, Dublin 2, Ireland (a division of Penguin Books Ltd.) • Penguin Group (Australia), 250 Camberwell Road, Camberwell, Victoria 3124, Australia (a division of Pearson Australia Group Pty. Ltd.) • Penguin Books India Pvt. Ltd., 11 Community Centre, Panchsheel Park, New Delhi—110 017, India • Penguin Group (NZ), 67 Apollo Drive, Rosedale, North Shore, Auckland 1311, New Zealand (a division of Pearson New Zealand Ltd.) • Penguin Books (South Africa) (Pty.) Ltd., 24 Sturdee Avenue, Rosebank, Johannesburg 2196, South Africa • Penguin Books Ltd., Registered Offices: 80 Strand, London WC2R 0RL, England

Copyright © 2014 by Penguin Group (USA) Inc.

IDIOT'S GUIDES and Design are trademarks of Penguin Group (USA) Inc.

International Standard Book Number: 978-1-61564-500-8
Library of Congress Catalog Card Number: 2013957749

16 15 14   8 7 6 5 4 3 2 1

Interpretation of the printing code: The rightmost number of the first series of numbers is the year of the book's printing; the rightmost number of the second series of numbers is the number of the book's printing. For example, a printing code of 14-1 shows that the first printing occurred in 2014.

Printed in the United States of America

Note: This publication contains the opinions and ideas of its author. It is intended to provide helpful and informative material on the subject matter covered. It is sold with the understanding that the author and publisher are not engaged in rendering professional services in the book. If the reader requires personal assistance or advice, a competent professional should be consulted.

The author and publisher specifically disclaim any responsibility for any liability, loss, or risk, personal or otherwise, which is incurred as a consequence, directly or indirectly, of the use and application of any of the contents of this book.

Publisher: *Mike Sanders*
Executive Managing Editor: *Billy Fields*
Senior Acquisitions Editor: *Tom Stevens*
Development Editor: *Kayla Dugger*
Senior Production Editor: *Janette Lynn*

Cover Designer: *Laura Merriman*
Book Designer: *William Thomas*
Indexer: *Tonya Heard*
Layout: *Brian Massey*
Proofreader: *Krista Hansing*

Contents

Introduction

Geometry requires a great deal of spatial sense. The ability to imagine figures in varied dimensions and transform these figures is not easy for all. Geometry also requires logical reasoning. The ability to use definitions and properties to validate conclusions can be a daunting task. As you sit in class or work on geometry problems, you may feel lost in the forest with no end in sight. Definitions sound like birds chirping and examples appear to be monkeys swinging from tree to tree. Don't worry, you are not alone!

We all have strengths and areas in need of improvement. What can help you through is perseverance and the right resource—for you, it's *Idiot's Guides: Geometry.*

In this book, we have presented geometry through illustrated examples and provided helpful hints to make geometry easy to understand. So rather than paying a personal tutor loads of money, let this book guide your way home through the forest of geometry!

How This Book Is Organized

This book is presented in seven parts:

In **Part 1, The Foundations of Geometry,** you will learn about all the basic elements of geometry. You will define, identify, and name points, lines, planes, and angles. You will also learn about the relationships formed between lines and angles and how to find their measures.

The theoretical part of geometry occurs in **Part 2, Reasoning and Proof.** You will learn the basics of writing proofs to form the basis for more in-depth proofs later in the book.

In **Part 3, Triangles,** you will see just how intricate these simple shapes can be. You will classify angles and find measures in various types of triangles. The proofs you learned about in the previous part will be extended here to prove that given triangles are congruent.

In **Part 4, Two-Dimensional Figures,** focuses on polygons and similarity. You will define, identify, and prove figures to be quadrilaterals (four-sided polygons). You will also look at relationships between sides and angles in polygons and learn how to prove figures are similar.

In **Part 5, Three-Dimensional Figures,** you explore three-dimensional figures (solids) and find out how they're related to their two-dimensional nets. You will also learn to find the area and volume of solid figures.

In **Part 6, Circles,** you will learn about the many lines, segments, and angles related to circles. You will also learn how to find measurements in circles as well as parts of circles.

Finally, in **Part 7, Coordinate Geometry,** you will learn to attack geometry problems that occur on a coordinate plane. This will include working with figures moving around on a coordinate plane, as well as writing equations for figures created by the intersection of a plane and a cone.

Extras

As teachers, there is so much wisdom we want to pass along to readers of this book. When we focus on a problem, sometimes we think about hints or cautions we would like to share with you. These sidebars are those little tidbits we've given to help you along the way.

DEFINITION

In these sidebars, you'll find definitions for common geometry terms.

HELPFUL POINT

Although we give you lots of information and examples in the text, from time to time there is a special piece of information we want to point out. These sidebars give you tips and thoughts on certain aspects of geometry.

IT DOESN'T ADD UP

With years of experience, we know some common errors that a lot of students make that we do not want you to make. These sidebars help you avoid these common pitfalls.

Acknowledgments

Being an author is a job that takes the time and patience of the not only authors themselves, but also the family and friends around them. With that said, we need to thank our husbands, Dharm and Troy, for taking over as father and mother to our wonderful children, Jake, Arjun, and Lily. Long hours in front of a computer were never questioned and meltdowns were handled with a sympathetic ear. If not for the support of these great men, this book would not have been possible.

Students are why we teach. Over the years, they have taught us just as much as we have taught them. We see mathematics through their eyes and it helps us to explain concepts in various ways. The opportunities we have had as instructors are valuable and countless. To our thousands of students over the years, *thank you!* Now, through this book, we get to teach thousands more.

Although we are teachers first and foremost, we always had a drive to be authors. In our practice, we have developed effective strategies to help students. Authoring our own books allows us to bring these strategies to more learners. This dream of authorship became reality through Mel Friedman. We thank him for always thinking of us and supporting us.

Now it is time to get a bit silly. Rebecca wants to thank Sonal, and Sonal wants to thank Rebecca. We are friends first and colleagues second. Our journey in education began over 10 years ago, and we motivate each other to be the best educators we can be. While we focused on separate parts of the book, we always reviewed and edited each other's work. We hope you finally understand geometry and become motivated to learn and understand some more.

We also want to dedicate this book to our moms, Bernadette (Rebecca's mom) and Gita (Sonal's mom). Without them, we would have never grown into the strong and intelligent women we are today. Although Bernadette is not with me anymore, her words constantly echo in my mind. She is the force that makes me who I am, and I miss her every day. Gita has spent her life encouraging my sister and me to reach for the stars, despite hard times like the passing of my father, Vikram, whose guidance from up above has also been a constant source of strength.

In closing, we give additional thanks to Tom Stevens and Kayla Dugger for their guidance through this process. They have more patience than you can imagine. Also, thanks to our extended family and friends, especially Sejal, Darshan, Minerva, Deepak, Don, Ann, Melissa, Jules, and Lois.

Trademarks

All terms mentioned in this book that are known to be or are suspected of being trademarks or service marks have been appropriately capitalized. Alpha Books and Penguin Group (USA) Inc. cannot attest to the accuracy of this information. Use of a term in this book should not be regarded as affecting the validity of any trademark or service mark.

The Foundations of Geometry

Mathematics is its own language. To study and understand geometry, many terms and basic elements need to be defined and explained. In this part, you learn about all the basic elements of geometry. Beginning with points, lines, and planes, you start to see how the structure of geometry is built and gain all the tools necessary to expand your learning of geometry into deeper concepts.

The Basics of Geometry

Did you ever try to define a word and circle back to the word you were defining? For example, what does *up* mean? I know what you are thinking—upward, up the stairs, up high in the sky, the opposite of down. It's common to use the exact word you are defining to define what it means. In fact, people need undefined terms to lay the foundation for what other terms mean.

Geometry is no different. Euclidean geometry is based on defined terms, undefined terms, and postulates that are used to prove many different theorems. In this chapter, we help you better understand those basic pieces of geometry.

In This Chapter

- Understanding points, lines, and planes
- Using symbols to identify lines, segments, and rays
- Learning about postulates involving points, lines, and planes

Basic Terminology

Words that you quite often use in everyday speech have different meanings in geometry. Let's take a look at some undefined terms and basic definitions.

Points, Lines, and Planes

Get a blank sheet of paper and pencil and draw two dots on the paper. In geometry, these dots are known as *points*. A point is an exact location, usually represented by a dot. A point has no dimension.

Now use a ruler to draw a straight figure that connects these dots and extends to the edges of the paper. The figure you drew to connect these dots is a *line*. A line is a straight geometric figure. It has no thickness, extends infinitely in both directions, and is one-dimensional.

The paper represents the *plane*. While a plane appears to have edges, it actually extends infinitely.

To think about these terms another way, look at Figure 1.1. The dot represents point *C,* the straight figure with arrows on both ends represents line *CD,* and the shape that looks like a piece of paper represents plane *M*.

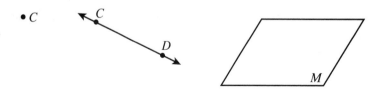

Figure 1.1: *Examples of a point, a line, and a plane.*

Segments and Rays

To visualize segments and rays, suppose you are a point on a map. Your location is Chicago, Illinois, and you are told to fly to Los Angeles, California. Your path can be represented by a *segment*. A segment is part of a line that does not extend indefinitely in either direction. The points on the map representing these cities are the *endpoints* of the segment.

 DEFINITION

Endpoints are points that indicate where a line begins and ends.

On the other hand, if you are simply told to fly east of Chicago with no destination, your path can be represented by a *ray*. A ray is part of a line that starts at a point and extends infinitely in one direction. In terms of this example, your initial point on the map of Chicago would be represented as an endpoint, and you would extend a line in one direction, east of where you started. If we started at Chicago and were told to fly west, the ray representing our path would be the opposite of the ray representing your path.

Now take a look at Figure 1.2. The straight figure with endpoints *A* and *B* represents segment *AB,* while the straight figure with endpoint *C* extending in the direction of *D* represents ray *CD.* Point *F* lies on a line between *E* and *G,* so ray *FE* and ray *FG* are opposite rays, or rays that share an endpoint yet extend in opposing directions.

Figure 1.2: *Examples of segments and rays.*

Identifying Parts of Geometric Figures

Now that you have a handle on the vocabulary related to points, lines, and planes, let's talk about the symbols that are used to identify the parts in Figure 1.3.

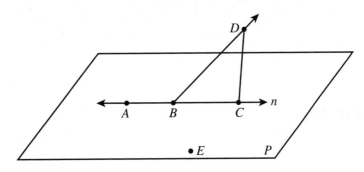

Figure 1.3: *Can you identify the parts of this geometric figure?*

To identify a line, you use two points on the line to name it. In Figure 1.3, *AC* is an example of a line. The mathematical symbol to denote a line is to write the letters *AC* with the image of a line over it, like so: $\overleftrightarrow{AC}$. You can also refer to that line as $\overleftrightarrow{CA}$.

You can identify a plane by the letter on the plane (*P* in Figure 1.3) or points on the plane. In order to use points to identify a plane, you must choose at least three points that lie on the plane and do not share a line, known as *coplanar points*. For example, plane *BEC* is another name for plane *P*. However, you would not be allowed to use points *A, B,* and *C* to name the plane because they are *collinear,* meaning the points lie on the same line.

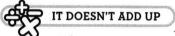

DEFINITION

If you were asked to stand in a line with other people, you and the others on this line would represent **collinear points.** Collinear points are points that lie on the same line. If you and these people were then asked to spread out, you and the others would represent **coplanar points.** Coplanar points are points that lie on the same plane.

An example of a segment in Figure 1.3 is segment *BC*, written $\overline{BC}$, with endpoints *B* and *C*. As long as you identify part of a line that ends on both sides, you are good to go! For this reason, $\overline{CB}$ is also a valid answer.

An example of a ray in Figure 1.3 is ray *BD*, written $\overrightarrow{BD}$, with an initial point (or endpoint) *B* that extends in the direction of *D*. $\overrightarrow{BC}$ and $\overrightarrow{BA}$ are opposite rays because they share an endpoint yet extend in opposing directions.

IT DOESN'T ADD UP

When it comes to rays, $\overrightarrow{BD}$ is not the same as $\overrightarrow{DB}$. For $\overrightarrow{BD}$, *B* is the endpoint and travels in the direction of *D*. For $\overrightarrow{DB}$, *D* is the endpoint and travels in the direction of *B*. Simply stated, traveling east from a location is not the same as traveling west from that same location.

Now that you have a handle on some basic definitions and the symbols associated with these terms, let's review them using Figure 1.4.

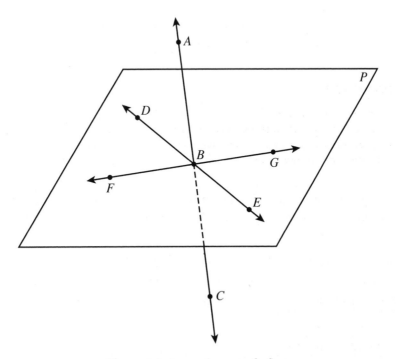

Figure 1.4: *Interpreting parts of a figure.*

As you can see, *D*, *B*, and *E* are collinear because they lie on the same line. *D*, *F*, *B*, and *G* are coplanar because they lie on the same plane. *A* is not coplanar to *F*, *B*, and *D*, because *A* does not lie on the same plane as *F*, *B*, and *D*. Finally, $\overrightarrow{BA}$ and $\overrightarrow{BC}$ are opposite rays because they share a common endpoint *B* and extend in opposing directions.

Relationships Between Lines and Planes

In the real world, points, lines, and planes work together to create the objects around you. Take a look at the chair or couch you are sitting on as you are reading this book. The legs of the chair or couch are lines that intersect the plane (that is, the seat) of it.

In geometry, the common words used to explain this type of relationship between points, lines, and planes are *intersecting, parallel,* and *skew.*

Intersect means to "cross or meet." The intersection of a line can be a point, while the intersection of a plane and another plane can be a line. For example, in Figure 1.5, $\overleftrightarrow{AC}$ intersects plane *P* at point *B* while plane *P* intersects plane *Q* at $\overleftrightarrow{XY}$.

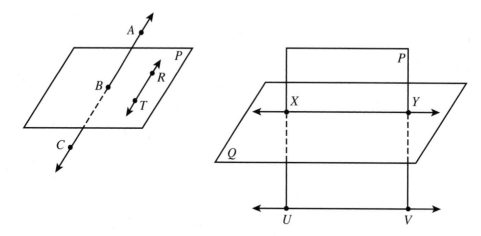

Figure 1.5: *Some relationships between lines and planes.*

When two lines do not cross and are coplanar, they are *parallel*. For example, in Figure 1.5, $\overleftrightarrow{XY}$ and $\overleftrightarrow{UV}$ are parallel.

When two lines do not intersect and are not coplanar, they are *skew*. For example, in Figure 1.5, $\overleftrightarrow{AC}$ and $\overleftrightarrow{RT}$ are skew.

HELPFUL POINT

How would you draw a line and a plane so the intersection is the line? You draw it so the entire line lies on the plane.

Now that you have a handle on the relationships between lines and planes, let's identify some relationships given Figure 1.6.

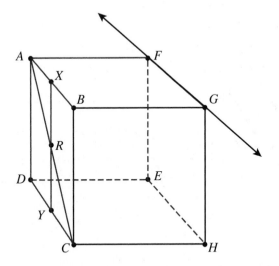

Figure 1.6: *Identifying relationships between lines and planes.*

Plane *ABCD* is parallel to plane *EFGH* and intersects plane *ABGF* at $\overleftrightarrow{AB}$. $\overleftrightarrow{FG}$ is the intersection of plane *EFGH* and plane *ABGF*. Also, $\overline{BG}$ is parallel to $\overleftrightarrow{CH}$; they share a plane and the segments will never intersect. $\overline{XY}$ and $\overleftrightarrow{FG}$ are skew because they will not intersect and do not share a plane.

Point, Line, and Plane Postulates

Just as terms like *up* are accepted without definition and are instead used to define other words, *postulates* are statements accepted without proof. Postulates are used to prove other rules, known as *theorems*.

DEFINITION

A **theorem** is a statement that can be proven using definitions and postulates.

Let's take a look at a basic example. In order to solve an equation such as $x - 5 = 13$, you add 5 to both sides:

$$x - 5 = 13$$

$$x - 5 \, (+ \, 5) = 13 \, (+ \, 5)$$

$$x = 18$$

You used the addition property of equality to prove $x = 18$ given $x - 5 = 13$. The addition property of equality is one of many algebraic postulates.

Similarly, basic geometry postulates serve as a starting point for proving other statements. The following are postulates you should know for points, lines, and planes:

- **Postulate 1: Two points determine a line.** For example, this can be used to prove the statement, "Two points always lie on one line."

- **Postulate 2: Three noncollinear points determine a plane.** For example, this can be used to prove the statement, "Three collinear points never determine a plane."

- **Postulate 3: If two points lie on a plane, then the line containing them lies on the plane.** For example, this can be used to prove the statement, "If two points are not on a plane, their line isn't on the plane."

- **Postulate 4: If two planes intersect, then their intersection is a line.** For example, this can be used to prove the statement, "If two planes do not intersect, they are parallel to one another."

Think about postulates this way. If you are driving from point A to point B, it is assumed that you can start the car. If you can't start the car, it is assumed that you can't drive from point A to point B. You must start somewhere, and in geometry, the starting point is a postulate!

The Least You Need to Know

- Conceptually understanding points, lines, and planes serves as the foundation to other concepts in geometry.
- Mathematic symbols are used to name lines, segments, and rays.
- Postulates are the starting points to prove other statements.

CHAPTER 2

Angles

When you park your car, sometimes you have a lot of room to park, and other times it is a tight squeeze. Did you realize that as you are thinking about how to the park your car, you are actually thinking about the *angle* you should take to turn into the spot?

There are angles all around you. Chairs are made for sitting, tables are made for dining, and houses are made to live in all because of the angle measures between the objects used to create other objects. In this chapter, we explore what an angle is, as well as identify various angle measures.

In This Chapter

- What are angles?
- Classifying angles based on their angle measure
- Using a protractor to determine angle measures
- Angle relationships and how to use them to solve equations

An Introduction to Angles

An *angle* is a figure formed by two rays that have the same endpoint. The common endpoint is the *vertex* of the angle. In Figure 2.1, the sides of the angle are $\overrightarrow{YX}$ and $\overrightarrow{YZ}$ and the vertex is point *Y*.

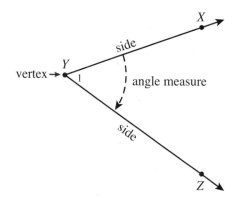

Figure 2.1: *Identifying parts of an angle.*

There are multiple ways to name an angle. The symbol $\angle$ refers to an angle. The angle in Figure 2.1 can be called $\angle\, Y$, $\angle\, XYZ$, $\angle\, ZYX$, or $\angle\, 1$.

HELPFUL POINT

When you name angles using three letters, the vertex point is always in the middle.

Let's look at another example using Figure 2.2. Each of the angles can be named by letters or numbers. Can you give both of their names?

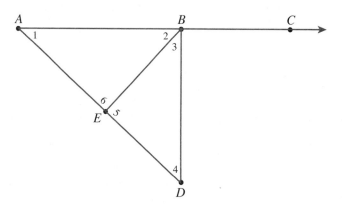

Figure 2.2: *Naming the angles.*

Some name-number pairings for the angles in Figure 2.2 are $\angle ABE$ and $\angle 2$, $\angle BED$ and $\angle 5$, and $\angle AEB$ and $\angle 6$.

Types of Angles

Angle measures created by lines and segments are not the same. Using Figure 2.3, let's name different types of angles and identify examples from this figure.

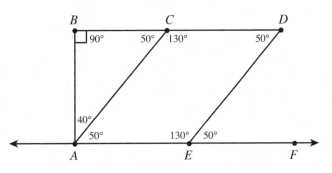

Figure 2.3: *Visualizing different angle measures.*

- An angle that measures greater than 0° but less than 90° is an *acute angle*. In Figure 2.3, $\angle BCA$, $\angle BAC$, $\angle CAE$, $\angle CDE$, and $\angle DEF$ are acute angles.

- An angle that measures greater than 90° but less than 180° is an *obtuse angle*. In Figure 2.3, $\angle ACD$ and $\angle AED$ are obtuse angles.

- An angle that measures 90° is a *right angle*. In Figure 2.3, $\angle B$ is a right angle. Note the symbol used to mark a right angle.

- An angle that measures 180° is a *straight angle*. In Figure 2.3, $\angle AEF$ is a straight angle. Notice how it forms a line and its sides $\overrightarrow{EA}$ and $\overrightarrow{EF}$ form opposite rays.

How to Measure Angles

Just as you could use a ruler to measure the length of a side in inches, you can use a *protractor* to determine an angle measure in degrees. The mathematical symbol to indicate a degree measure is °.

In Figure 2.4, you can see how to use a protractor to measure $\angle AVB$. To measure this angle, you align vertex V at the *midpoint* on the straightedge of the protractor.

DEFINITION

A **protractor** is a tool used to measure the rotation of one ray away from the other about the vertex point. A **midpoint** is the middle of a segment.

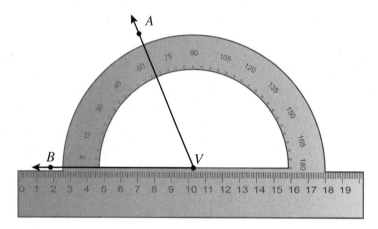

Figure 2.4: *Measuring an angle using a protractor.*

Notice how the straightedge of the protractor aligns to $\overrightarrow{VB}$ at 0°. The measure of $\angle AVB$ is the measure at which $\overrightarrow{VA}$ passes through the protractor. Therefore, the measure of $\angle AVB$, denoted "$m\angle AVB$," is 70°.

In Figure 2.4, we used a protractor to determine angle measures. See if you can do the same for $\angle DOG$, $\angle FOG$, $\angle GOT$, and $\angle HOG$ and name their angle measures and types.

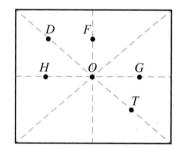

Figure 2.5: *Finding angle measures.*

The $m\angle DOG$ is 135°. Because the angle is greater than 90° and less than 180°, it is an obtuse angle. The $\angle FOG$ is 90°. Because the angle is exactly 90°, it is a right angle. The $\angle GOT$ is 45°. Because the angle is less than 90° and greater than 0°, it is an acute angle. The $\angle HOG$ is 180°. Because the angle is a line, it is a straight angle. See? That's not so difficult!

Angle Relationships

In geometry, certain pairs of angles have special relationships. These relationships are useful in determining unknown angle measures. The following sections discuss some of these angle pair relationships.

Complementary and Supplementary Angles

When the sum of the measurements of two angles equals 90°, the angles are considered *complementary*. When the sum of two angles is 180°, the two angles are considered *supplementary*.

Let's take a look at the drawing of a bathroom. If you are standing inside and open the door so that it creates a 60° angle with the exit, how many more degrees would you have to open the door so the angle between the exit and the bathroom wall is 90°?

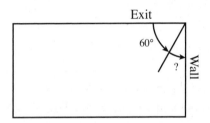

Figure 2.6: *Angle to open a bathroom door.*

If you said 30°, you are absolutely correct! When you are opening and closing doors that open to a 90° angle, the angles formed on either side of the door are complementary.

Now say the wall lined up with the exit; in this case, the door would open to a 180° angle. You would then have to open the door 120° so that angle between the exit and the bathroom wall would be 180°. When the door opens so that the angles form a semicircle, the angles formed on either side of the door are supplementary.

 DEFINITION

Complementary angles are two angles whose measures add up to exactly 90°. **Supplementary angles** are two angles whose measures add up to exactly 180°.

Let's take a look at another example. In Figure 2.7, can you identify which angle pair is complementary and which angle pair is supplementary?

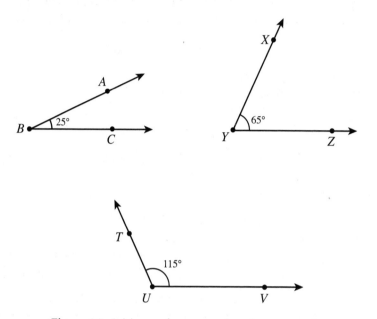

Figure 2.7: *Pairing complementary and supplementary angles.*

In the figure, $\angle ABC$ and $\angle XYZ$ are complementary angles:

$m\angle ABC + m\angle XYZ$

$= 25° + 65°$

$= 90°$

Because they add up to 90°, you know for sure these are complementary.

And you can probably guess that $\angle XYZ$ and $\angle TUV$ are supplementary. Let's double-check, though:

$m\angle XYZ + m\angle TUV$

$= 65° + 115°$

$= 180°$

As you can see, the two angles add up to 180°, meaning they're supplementary.

Adjacent and Vertical Angles

Two angles that have a common vertex and a common side are *adjacent angles*. In Figure 2.8, there are four adjacent pairs of angles: $\angle 1$ and $\angle 2$, $\angle 2$ and $\angle 3$, $\angle 3$ and $\angle 4$, and $\angle 1$ and $\angle 4$.

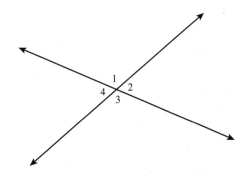

Figure 2.8: *Examples of adjacent and vertical angles.*

These pairs are also known as *linear pairs*. Angles that form a linear pair are adjacent as well as supplementary. As you can see, these adjacent angles add up to 180° and form a line.

Another relationship between two angles you should know about is *vertical angles*. These are angles that don't share a side and have the same angle measure. In Figure 2.8, $\angle 1$ and $\angle 3$ are an example of vertical angles.

> **DEFINITION**
>
> **Vertical angles** are formed by two intersecting lines. Vertical angles do not share a common side and have the same measure.

Can you identify another pair of vertical angles in Figure 2.8? If you said $\angle 2$ and $\angle 4$, you are correct!

Now that you have a handle on angle measures and angle pair relationships, let's take a look at Figure 2.9. What are some of the angle pair relationships based on the angle measures?

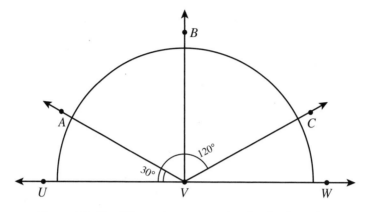

Figure 2.9: *Identifying angle measures and angle pair relationships.*

For example, the $m\angle UVA = 30°$ and the $m\angle AVB = 60°$. Because the $m\angle UVA$ plus the $m\angle AVB$ is 90°, $\angle UVA$ and $\angle AVB$ are complementary angles. The $m\angle UVC = 120°$ and $m\angle CVW = 30°$. Because $m\angle BVC$ plus $m\angle CVW$ is 180°, $\angle UVC$ and $\angle CVW$ are supplementary angles. Finally, $\angle UVA$ and $\angle AVW$ form a linear pair because the angles are adjacent and the sum of the angles measures is 180°.

Using Angle Relationships to Solve Equations

Sometimes measures of angles are written as expressions. You will need to use these expressions to write an equation based on the angle pair relationship.

For example, in Figure 2.10, can you identify the values of *r*, *s*, and *t*?

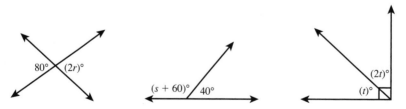

Figure 2.10: *Using angle relationships to find the values of r, s, and t.*

For the figure containing r, the expressions provided are of vertical angles. Because vertical angles have the same measure, you must set the expressions equal to one another and then solve for r:

$$2r = 80$$
$$2r \div 2 = 80 \div 2$$
$$r = 40$$

You can now verify your answer by putting 40 in for r:

$$2r = 2(40) = 80$$

 **HELPFUL POINT**

Don't forget to check your work by substituting the answer in for the variable.

For the figure containing s, the expressions provided are of a linear pair of angles. Because the sum of the measures of a linear pair of angles must equal 180°, you must set the sum of expressions equal to 180 and solve for s:

$$s + 60 + 40 = 180$$
$$s + 100 = 180$$
$$s + 100 - 100 = 180 - 100$$
$$s = 80$$

You can check your answer:

$$s + 60 + 40 = 180$$
$$80 + 60 + 40 = 180$$

Bravo!

For the figure containing t, the expressions provided are of a pair of complementary angles. Because the sum of these angles must be 90°, you must set the sum of expressions equal to 90 and then solve for t:

$$2t + t = 90$$
$$3t = 90$$
$$3t \div 3 = 90 \div 3$$
$$t = 30$$

Feel free to verify your answer:

2(30) + (30) = 90

3(30) = 90

The Least You Need to Know

- An angle is made up of two rays with the same endpoint, or vertex.
- Angles are classified as acute, obtuse, right, and straight based on their measures.
- You can use angle relationships to determine missing angle measures in expressions.

CHAPTER 3

Segment and Angle Addition

Quite often, ideas related to simple arithmetic, such as adding and subtracting, can be used to interpret a geometric concept. In this chapter, we discuss segment addition as related to the Segment Addition Postulate and angle addition as related to the Angle Addition Postulate.

Segment Addition

Because line segments have endpoints (see Chapter 1), they are able to be measured. Think about what would happen if two segments were connected. If you know the length of each of the segments, you can add them together to find the length of the combined segment. This is known as the *Segment Addition Postulate.*

In This Chapter

- Adding and finding lengths of segments
- Adding and finding angle measures
- When segments or angles are congruent

 **DEFINITION**

The **Segment Addition Postulate** states that if B is between A and C, then $AB + BC = AC$.

Numerical Examples

In Figure 3.1, two segments, $\overline{AB}$ and $\overline{BC}$, are joined together to make a longer segment. If $AB =$ 7 inches and $BC = 13$ inches, what is the length of AC?

Figure 3.1: *Adding two segments, $\overline{AB}$ and $\overline{BC}$.*

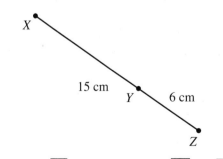 **HELPFUL POINT**

Another way of denoting the length of $\overline{AC}$ is to simply write AC without the line above it.

When looking at the figure, the larger segment, $\overline{AC}$, is split into two smaller segments, $\overline{AB}$ and $\overline{BC}$. Therefore, according to the Segment Addition Postulate, $AB + BC = AC$.

Now that you know the equation, you can substitute the values into it to find AC:

$7 + 13 = AC$

$20 \text{ inches} = AC$

But what do you do when you have one of the parts and the whole? In Figure 3.2, you see two segments, $\overline{XY}$ and $\overline{YZ}$, that combine to create $\overline{XZ}$. If $XZ = 15$ cm and $YZ = 6$ cm, what is the length of XY?

Figure 3.2: $\overline{XZ}$ *with smaller segments* $\overline{XY}$ *and* $\overline{YZ}$.

According to the Segment Addition Postulate, $XY + YZ = XZ$. Substitute the values into the equation:

$$XY + 6 = 15$$

In this case, you have to subtract the smaller value from each side to get the answer:

$$XY + 6 - 6 = 15 - 6$$

$$XY = 9 \text{ cm}$$

In some cases, a segment may have a midpoint. From Chapter 2, you may recall that a midpoint is a point on a segment that is halfway between the endpoints. Therefore, the two smaller segments formed are the same length, or *congruent*. The symbol for congruency is $\cong$.

In Figure 3.3, M is the midpoint. If $LN = 20$ cm, what are the lengths of LM and MN?

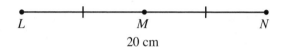

$$L \qquad\qquad M \qquad\qquad N$$

20 cm

Figure 3.3: $\overline{LN}$ *with two congruent segments,* $\overline{LM}$ *and* $\overline{MN}$.

HELPFUL POINT

The vertical markings on the segments in Figure 3.3 indicate they are congruent.

Because M is the midpoint and creates two congruent segments, $\overline{LM}$ and $\overline{MN}$ must have the same length. Therefore, divide the length of $\overline{LN}$ by 2 to get the length of each of the segments.

$\overline{MN}$ is found the same way and has the same length of 10 centimeters.

Algebraic Examples

Now that you have a handle on segment addition, let's mix in some more algebra. It is possible that the lengths of segments will be written as expressions. In such cases, you will need to use the expressions to write an equation. The equation can then be used to solve for the actual segment lengths.

Figure 3.4 provides expressions for $\overline{AB}$ and $\overline{BC}$. If $AC = 29$ inches, what are the lengths of AB and BC?

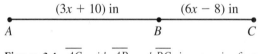

Figure 3.4: $\overline{AC}$ *with* $\overline{AB}$ *and* $\overline{BC}$ *in expression form.*

In the figure, $AB + BC = AC$, so substitute the expressions into the equation and simplify:

$(3x + 10) + (6x - 8) = 29$

$(3x + 6x) + (10 - 8) = 29$

$9x + 2 = 29$

Now solve the equation for x:

$9x + 2 = 29$

$9x + 2 - 2 = 29 - 2$

$9x = 27$

$\dfrac{9x}{9} = \dfrac{27}{9}$

$x = 3$

The value of $x = 3$ now needs to be substituted into the original expressions to find the actual length of each segment. The following shows you how to put the number in for one of the segments:

$AB = (3x + 10)$

$AB = [3(3) + 10]$

$AB = (9 + 10)$

$AB = 19$ inches

You can either complete the same process for $\overline{BC}$ or you can subtract the part found, $AB = 19$ inches, from the whole, $AC = 29$ inches, to determine the length of the other part. In the equation $AB + BC = 29$, you substitute 19 for AB:

$19 + BC = 29$

$19 - 19 + BC = 29 - 19$

$BC = 10$ inches

For some segment addition equations, the full segment length might be an expression. In Figure 3.5, EF is 14 yards, FG is $4x$ yards, and $EG = 5x + 12$ yards. What are the lengths of FG and EG?

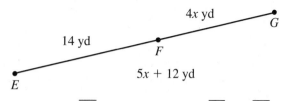

Figure 3.5: $\overline{EG}$ *in expression form with* $\overline{EF}$ *and* $\overline{FG}$

HELPFUL POINT

When information is given, it is beneficial to label the figure with that information like you see throughout this chapter. That way, you immediately know what to plug in the Segment Addition Postulate setup.

For the figure given, $EF + FG = EG$, so substitute the given information into the equation:

$14 + 4x = 5x + 12$

Now solve the equation for x:

$14 + 4x - 4x = 5x - 4x + 12$

$14 = x + 12$

$14 - 12 = x + 12 - 12$

$2 = x$

The value of $x = 2$ now needs to be substituted into the original expressions. The following shows how to do so for FG:

$FG = 4x$

$FG = 4 \times 2$

$FG = 8$ yards

Because $EF = 14$ yards and $FG = 8$ yards, their sum is the length of EG:

$EG = 14 + 8$

$EG = 22$ yards

Angle Addition

Angles are formed by two rays with the same endpoint. Just as in segments, angles can be added together to form a larger angle. This is called the *Angle Addition Postulate* and occurs when two angles are adjacent. If you recall from Chapter 2, adjacent angles are angles that share a common vertex and a common ray.

Numerical Examples

Let's take a look at an example. In Figure 3.6, $m\angle ABC = 30°$ and $m\angle CBD = 40°$. What is the measure of $\angle ABD$?

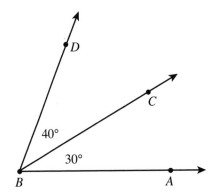

Figure 3.6: *Two adjacent angles combine to form one larger angle.*

A pair of adjacent angles, $\angle ABC$ and $\angle CBD$, combine to create a larger angle, $\angle ABD$. $\angle ABC$ and $\angle CBD$ are adjacent because they share $\overline{BC}$ and have a common vertex. Therefore, $m\angle ABC + m\angle CBD = m\angle ABD$. Substitute the values into the equation to get the answer:

$$30° + 40° = m\angle ABD$$

$$70° = m\angle ABD$$

IT DOESN'T ADD UP

Remember that when you name angles, the vertex point is always in the middle. Don't let the vertex point letter fall anywhere else.

In the preceding example, the measures of two smaller adjacent angles were given and needed to be added together to find the measure of the larger angle. In the following example, though, we are going to show you what to do when the measure of one of the adjacent angles is missing.

In Figure 3.7, $m\angle WXY = 63°$ and $m\angle WXZ = 94°$. What is the measure of $m\angle YXZ$?

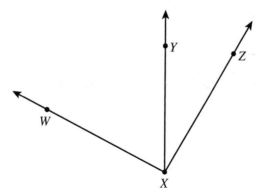

Figure 3.7: *A pair of adjacent angles, $\angle YXZ$ and $\angle WXZ$, combine to create a larger angle, $m\angle WXY$.*

Using angle addition, $m\angle WXY + m\angle YXZ = m\angle WXZ$, so substitute the values into the equation:

$$63° + m\angle YXZ = 94°$$

$$63° - 63° + m\angle YXZ = 94° - 63°$$

$$m\angle YXZ = 31°$$

Before moving on to the algebraic examples, let's talk about *angle bisectors*. An angle bisector is a ray that separates an angle into two congruent angles.

In Figure 3.8, $\overrightarrow{RS}$ is an angle bisector. If $m\angle QRT = 120°$, what are the measures of $m\angle QRS$ and $m\angle SRT$?

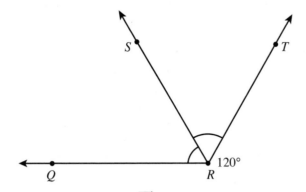

Figure 3.8: *The common ray,* $\overrightarrow{RS}$ *, is an angle bisector of the* $\angle$ *QRT.*

Because $\overrightarrow{RS}$ is an angle bisector of $\angle QRT$, that means $\angle QRS \cong \angle SRT$, which is shown in the figure by each angle having the same marking. Due to the angles being congruent, their measures are half of $m\angle QRT$: Divide 120° by 2 to get the angle measurements for $m\angle QRS$ and $m\angle SRT$:

$$\frac{120°}{2} = 60°$$

So $m\angle QRS$ and $m\angle SRT$ are each 60°.

Algebraic Examples

As with segment addition, angle addition can include algebraic expressions and equations. When expressions are given for the measure of angles, you need to set up and solve an equation to find the measurement.

In Figure 3.9, if $\angle ABC = (2x)°$ and $\angle CBD = (4x + 6)°$, what is the measure of $\angle ABD$?

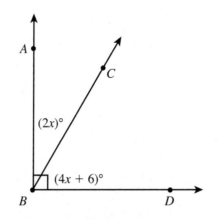

Figure 3.9: *The algebraic expressions (2x)° and (4x + 6)° indicate the measure of $\angle$ ABC and $\angle$ CBD, respectively.*

HELPFUL POINT

Do you see the little square by vertex B? That means $\angle ABD$ is a right angle and measures 90°.

Using the equation for angle addition, $m\angle ABC + m\angle CBD = m\angle ABD$, so substitute the values into the equation:

$$(2x) + (4x + 6) = 90$$

Solve the equation for x:

$$(2x + 4x) + (6) = 90$$

$$6x + 6 = 90$$

$$6x + 6 - 6 = 90 - 6$$

$$6x = 84$$

$$\frac{6x}{6} = \frac{84}{6}$$

$$x = 14$$

Now use $x = 14$ to substitute into the original expressions:

$$m\angle ABC = 2x = 2(14) = 28°$$

Because $\angle ABC$ is 28° and $\angle ABD$ is 90°, you can subtract these measures to find $m\angle CBD$:

$$90° - 28° = 62°$$

Let's look at one last example. In Figure 3.10, $\overrightarrow{GH}$ is a bisector of $\angle FGI$, so $m\angle FGH$ and $m\angle HGI$ are equal. If $m\angle FGI = 100°$ and $m\angle FGH = (2x+5)°$, what are the measures of $m\angle FGH$ and $m\angle HGI$?

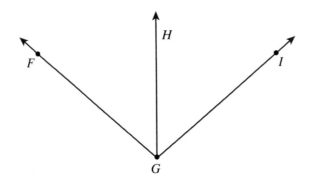

Figure 3.10: *A pair of adjacent angles, $m\angle FGH$ and $m\angle HGI$, combine to create a larger angle, $\angle FGI$.*

Because ray GH is a bisector of $\angle FGI$, that means $\angle FGH \cong \angle HGI$. With the two angles being congruent, you can say that $m\angle FGH$ and that $m\angle HGI$ is $(2x + 5)°$. Substitute the values into the equation:

$$(2x + 5) + (2x + 5) = 100$$

Solve the equation for x:

$$(2x + 2x) + (5 + 5) = 100$$

$$4x + 10 = 100$$

$$4x + 10 - 10 = 100 - 10$$

$$4x = 90$$

$$\frac{4x}{4} = \frac{90}{4}$$

$$x = 22.5$$

Substitute the $x = 22.5$ into the original expression:

$$(2x + 5) = 2(22.5) + 5 = 45 + 5 = 50°$$

Therefore, $m\angle FGH$ and $m\angle HGI$ are each 50°.

The Least You Need to Know

- The Segment Addition Postulate is used to find missing lengths of segments.
- The Angle Addition Postulate is used to find missing angle measures.
- Congruency occurs when two figures are the same shape and the same size.

CHAPTER 4

Introduction to Coordinate Geometry

Just think if your best friend lived all the way on the other side of town. What would be the best location for you two to meet? The midpoint of your two houses! So just how far does each of you have to walk? Could you figure this out? This is where coordinate geometry comes in.

In this chapter, we show you how to plot coordinates, find the midpoint and the distance between two points, and identify relationships between lines.

In This Chapter

- Plotting ordered pairs
- Figuring out the coordinates for points
- Finding the distance and the halfway point for two points
- Using slope to prove lines are parallel or perpendicular

Coordinate Planes

In Chapter 1, we told you that a plane is a flat two-dimensional surface that endlessly extends in all directions. If there was a dot on this plane, you could be searching for it for a very long time. That is why there are *coordinate planes*. Coordinate planes allow you to pinpoint locations on a plane.

The coordinate plane is the foundation of *coordinate geometry*. The horizontal number line is called the *x-axis*, the vertical number line is called the *y-axis*, and their intersection point is called the *origin*. The origin is located at the *ordered pair* (0, 0) because the value of *x* is 0 and the value of *y* is 0.

📖 **DEFINITION**

A **coordinate plane** is a plane formed by the intersection of a horizontal and vertical number line.

Coordinate geometry is an area of geometry where the position of points on the plane is given using an ordered pair of numbers.

An **ordered pair** is a pair of numbers used to locate a point on a coordinate plane. It is in the form (x, y).

When the x-axis and y-axis intersect, they section the plane into four parts, known as *quadrants*. I have labeled all the parts of the coordinate plane in Figure 4.1. The x-value of an ordered pair is positive in the first and fourth quadrants and negative in the second and third. The y-value of an ordered pair is positive in the first and second quadrants and negative in the third and fourth.

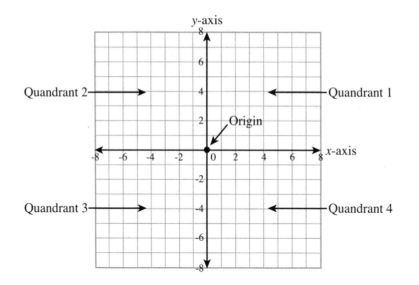

Figure 4.1: *A coordinate plane has four quadrants separated by the x-axis and y-axis.*

So how do you get these ordered pairs on the coordinate plane? Let's say you have to plot the ordered pair (4, -5). For the ordered pair (4, -5), the x-value is 4 and the y-value is -5. Because the x-value is positive, you will begin at the origin and move 4 spaces to the right. From that point, because the y-value is negative, you will move 5 spaces down. You can see the ordered pair plotted in Figure 4.2.

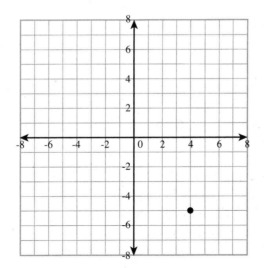

Figure 4.2: *The ordered pair (4, -5) is plotted on the coordinate plane.*

IT DOESN'T ADD UP

When plotting ordered pairs, do not confuse *x* and *y*. The *x*-axis goes right (positive) and left (negative), while the *y*-axis goes up (positive) and down (negative).

Now we are going to give you a coordinate plane with some points plotted, and you are going to find the ordered pair that names each of their locations. Did you ever hear the saying "X marks the spot"? For these, it's like knowing how many spaces to walk in each direction.

Figure 4.3 shows three points: *A, B,* and *C.* What are the coordinates?

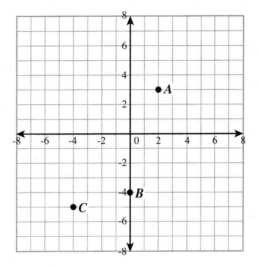

Figure 4.3: *Three points are plotted on the coordinate plane.*

Let's work with point *A* first. Whenever you are finding the ordered pair of a point, start at the origin. Starting at the origin, you must move 2 spaces to the right and 3 spaces up to reach point *A*. These are both positive movements, so you are at the ordered pair (2, 3)—these are the coordinates for point *A*.

Going back to the origin, you must move 0 spaces left or right and 4 spaces down to reach point *B*. Because there is downward movement, the *y*-value of the ordered pair will be negative. Therefore, point *B* is located at (0, -4).

Lastly, you must go back to the origin to find the location point *C*. From the origin, you must move left 4 spaces and down 5 spaces to reach point *C*. These are both negative movements, so the ordered pair for point *C* is (-4, -5).

Finding the Distance Between Two Points

When working with two points, you may sometimes have to find the distance between the points.

When points on a coordinate plane are on the same horizontal or vertical line, as in Figure 4.4, you can find the distance between the two points simply by counting the spaces.

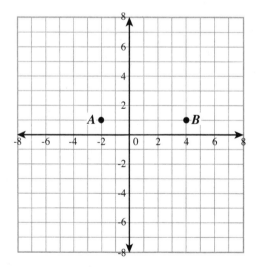

Figure 4.4: *Points A and B are on the same horizontal line.*

If you count the number of spaces it takes to get from point *A* to point *B*, you'll see that the distance between them is 6 units.

But what if the points are not on the same horizontal or vertical line? For these instances, you need to use the distance formula:

$$d = \sqrt{\left(x_2 - x_1\right)^2 + \left(y_2 - y_1\right)^2}$$

In the preceding formula, the x's and y's represent the numbers from two ordered pairs, (x_1, y_1) and (x_2, y_2). By substituting the numbers into the formula, you can find the distance between any two points on a coordinate plane.

> **HELPFUL POINT**
>
> The subscripts, or the tiny numbers next to the x and y, indicate the first and second points. It does not matter which point is first and which one is second—the answer will be the same.

To see how the distance formula works, let's find the distance between the points (0, -2) and (8, 4).

Using (0, -2) as (x_1, y_1) and (8, 4) as (x_2, y_2), substitute the numbers into the formula and solve:

$$d = \sqrt{\left(x_2 - x_1\right)^2 + \left(y_2 - y_1\right)^2}$$
$$d = \sqrt{\left(8 - 0\right)^2 + \left(4 - (-2)\right)^2}$$
$$d = \sqrt{\left(8\right)^2 + \left(6\right)^2}$$
$$d = \sqrt{64 + 36}$$
$$d = \sqrt{100} = 10$$

The distance between the points (0, -2) and (8, 4) is 10 units.

When finding the distance between two points on a coordinate plane, you are also finding the length of a segment that connects the two points.

For example, Figure 4.5 shows $\overline{AB}$ graphed on the coordinate plane. You can use the distance formula to find the length of the segment.

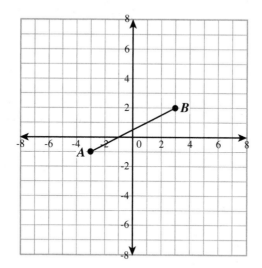

Figure 4.5: *Points A and B mark the endpoints of the segment.*

Before substituting into the distance formula, you need to correctly identify the coordinates of point A and point B. Point A is 3 units left of the origin and 1 unit down, which is the ordered pair (-3, -1). Point B is 3 units right of the origin and 2 units up, which is the ordered pair (3, 2).

Using (-3, -1) as (x_1, y_1) and (3, 2) as (x_2, y_2), substitute the values into the formula and solve:

$$d = \sqrt{\left(x_2 - x_1\right)^2 + \left(y_2 - y_1\right)^2}$$
$$d = \sqrt{\left(3 - (-3)\right)^2 + \left(2 - (-1)\right)^2}$$
$$d = \sqrt{\left(6\right)^2 + \left(3\right)^2}$$
$$d = \sqrt{36 + 9}$$
$$d = \sqrt{45} \approx 6.71$$

The distance between the points (-3, -1) and (3, 2) is approximately 6.71 units.

HELPFUL POINT

The squiggly equals sign means "approximately." Because $\sqrt{45}$ results in an ongoing decimal, you have to round the result, making the answer approximate instead of exact.

Midpoint

Do you remember from the introduction of this chapter where we talked about the best place to meet your friend? That's right—the midpoint between your two houses. The midpoint is another name for the halfway point.

Now that you know how to find the distance between two points, let's look at finding the midpoint.

Figure 4.6 should be familiar to you; it's what you used when finding the distance for two points on the same horizontal line. You know that the distance is 6 units; what's the location of the midpoint?

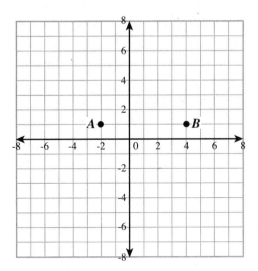

Figure 4.6: *Point A and point B are on the same horizontal line.*

Half of 6 spaces is 3 spaces, so if you count three spaces from either endpoint, the midpoint is located at point *M*, which is (1, 1).

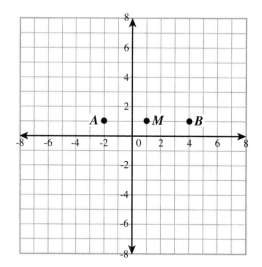

Figure 4.7: *Point M is the midpoint between point A and point B.*

What if the points aren't on the same line? For example, look at point *C* and point *D* in Figure 4.8. Because they're not on the same line, you can't just count the number of spaces. So what do you do?

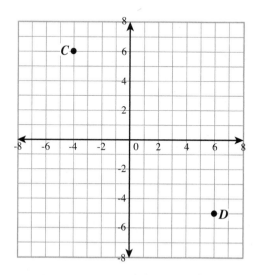

Figure 4.8: *Two points are plotted on the coordinate plane.*

For points that aren't on the same line, you need to use the midpoint formula:

$$M = \left(\frac{x_1 + x_2}{2}, \frac{y_1 + y_2}{2} \right)$$

Remember from the distance formula that the subscripts are just indicating the first ordered pair and the second ordered pair.

For this example, point C is located at (-4, 6) and point D is located at (6, -5). Using (-3, -1) as (x_1, y_1) and (3, 2) as (x_2, y_2), substitute the values into the formula and solve:

$$M = \left(\frac{x_1 + x_2}{2}, \frac{y_1 + y_2}{2} \right)$$

$$M = \left(\frac{-3 + 3}{2}, \frac{-6 + 2}{2} \right)$$

$$M = \left(\frac{0}{2}, \frac{-4}{2} \right)$$

$$M = (0, -2)$$

So the midpoint between points C and D is located at (0, -2).

HELPFUL POINT

Just to make sure you have this, try one more example: find the midpoint of a segment with endpoints (-4, -6) and (0, -8).

Using (-4, -6) as (x_1, y_1) and (0, -8) as (x_2, y_2), substitute into the formula:

$$M = \left(\frac{x_1 + x_2}{2}, \frac{y_1 + y_2}{2} \right)$$

$$M = \left(\frac{-4 + 0}{2}, \frac{-6 + (-8)}{2} \right)$$

$$M = \left(\frac{-4}{2}, \frac{-14}{2} \right)$$

$$M = (-2, -7)$$

The midpoint of a segment with endpoints (-4, -6) and (0, -8) is (-2, -7).

Things are now about to get a little more complicated. Sometimes you can be given the coordinates of the midpoint and one of the endpoints. What's missing? That's right, the other endpoint.

For example, on a coordinate plane, $\overline{XY}$ has endpoint Y at (8, 4). The midpoint of $\overline{XY}$ is (2, -1) and is labeled point M. What are the coordinates of point X?

First, identify the coordinates of point X, point Y, and the midpoint M:

$$X = (x, y) \qquad Y = (8, 4) \qquad M = (2, -1)$$

Because you do not know the coordinates of X, you can leave the ordered pair as (x, y).

Now, recall the midpoint formula. That's what you're going to use to find point X. Using (x, y) as (x_1, y_1), (8, 4) as (x_2, y_2), and (2, -1) as M, substitute the values into the formula:

$$M = \left(\frac{x_1 + x_2}{2}, \frac{y_1 + y_2}{2} \right)$$

$$(2, -1) = \left(\frac{x + 8}{2}, \frac{y + 4}{2} \right)$$

Notice that (2, -1) replaced the M (midpoint) and (8, 4) replaced x_2 and y_2. At this point, separate the equation into two new equations—one for the x-coordinate and one for the y-coordinate.

The equation to find the x-coordinate is as follows:

$$2 = \frac{x + 8}{2}$$

$$2 \cdot 2 = \left(\frac{x + 8}{2} \right) \cdot 2$$

$$4 = x + 8$$

$$4 - 8 = x + 8 - 8$$

$$-4 = x$$

The equation to find the y-coordinate is as follows:

$$-1 = \frac{y+4}{2}$$

$$-1 \cdot 2 = \left(\frac{y+4}{2}\right) \cdot 2$$

$$-2 = y + 4$$

$$-2 - 4 = y + 4 - 4$$

$$-6 = y$$

So the coordinate of point X is (-4, -6).

Linear Relationships

If you recall from Chapter 1, parallel lines are lines that never intersect. Perpendicular lines, on the other hand, are lines that intersect, forming a right angle. In this section, we are going to talk about these lines on the coordinate plane.

In Figure 4.9, the lines graphed on the first coordinate plane are parallel, while the lines graphed on the second coordinate plane are perpendicular.

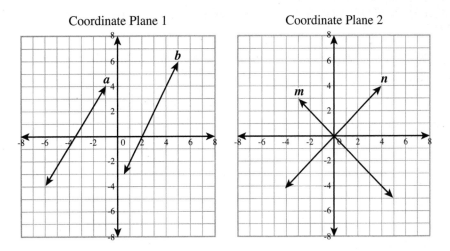

Figure 4.9: *Two parallel lines, a and b, are graphed on coordinate plane 1; two perpendicular lines, m and n, are graphed on coordinate plane 2.*

We have told you that the lines above are parallel and perpendicular, but are you just going to take our word for it? What if we give you another coordinate plane with two lines; how could you be sure they're parallel or perpendicular?

Think back to algebra when you learned about graphing lines. If two lines were parallel, they had the same slope; if two lines were perpendicular, their slopes were opposite reciprocals. These two facts are going to help you prove if lines are parallel, perpendicular, or neither.

> **HELPFUL POINT**
>
> $\dfrac{-3}{2}$ and $\dfrac{2}{3}$ are opposite reciprocals—their signs are different and the fraction flipped.

Proving Lines Are Parallel

Look back to the parallel lines on coordinate plane 1 in Figure 4.9. To prove that lines *a* and *b* are parallel, you must show they have the same slope. So first, let's have a refresher on slope.

Slope is a constant rate of change. The value of the slope tells you the steepness of the line. Slope is represented by a fraction; the change in the *y*-value (movement up or down) is in the numerator, and the change in the *x*-value (movement right or left) is in the denominator. The formula for slope is the following:

$$m = \frac{change\ in\ y}{change\ in\ x} = \frac{y_2 - y_1}{x_2 - x_1}$$

Slope is represented by the variable *m*. To use the formula, you need to have two points on the line; the subscripts in the formula are identifying which of the two points goes where.

Now, for line *a* and line *b* to be parallel, they must have the same slope. Identify two points on each line. First, find the coordinates for two points on line *a*. In this case, let's use (-4, -1) and (-3, 1). Using (-4, -1) as (x_1, y_1) and (-3, 1) as (x_2, y_2), substitute these values into the formula:

$$m = \frac{y_2 - y_1}{x_2 - x_1}$$

$$\frac{y_2 - y_1}{x_2 - x_1} = \frac{1 - (-1)}{-3 - (-4)}$$

$$\frac{1 - (-1)}{-3 - (-4)} = \frac{2}{1}$$

$$\frac{2}{1} = 2$$

The slope of line *a* is 2.

⊗ IT DOESN'T ADD UP

When substituting in the slope formula, it is important to watch the signs. Too often, answers are incorrect because of computational errors with signs.

To see if line *b* has the same slope, start by identifying two points on the line. In this case, you can go with (2, 0) and (3, 2). Using (2, 0) as (x_1, y_1) and (3, 2) as (x_2, y_2), substitute the values into the formula:

$$m = \frac{y_2 - y_1}{x_2 - x_1}$$

$$\frac{y_2 - y_1}{x_2 - x_1} = \frac{2 - 0}{3 - 2}$$

$$\frac{2}{1} = 2$$

The slope of line *b* is 2.

Because the slope of line *a* is the same as the slope of line *b*, the lines are proven to be parallel.

Proving Lines Are Perpendicular

Let's move on to coordinate plane 2 in Figure 4.9 and figure out how to prove line *m* is perpendicular to line *n*. To be perpendicular, their slopes need to be opposite reciprocals.

Like you did when trying to prove lines were perpendicular, you need to first find two points on each line. Starting with line *m*, you can use the points (-2, 2) and (0, 0). Using (-2, 2) as (x_1, y_1) and (0, 0) as (x_2, y_2), substitute the values into the formula:

$$m = \frac{y_2 - y_1}{x_2 - x_1}$$

$$\frac{y_2 - y_1}{x_2 - x_1} = \frac{0 - 2}{0 - (-2)}$$

$$\frac{0 - 2}{0 - (-2)} = \frac{-2}{2}$$

$$\frac{-2}{2} = -1$$

The slope of line *m* is -1.

Now you have to see if the slope of line n is the opposite reciprocal of -1. Using the points (2, 2) as (x_1, y_1) and (0, 0) as (x_2, y_2), substitute the values into the formula:

$$m = \frac{y_2 - y_1}{x_2 - x_1}$$

$$\frac{y_2 - y_1}{x_2 - x_1} = \frac{0-2}{0-2}$$

$$\frac{0-2}{0-2} = \frac{-2}{-2}$$

$$\frac{-2}{-2} = 1$$

The slope of line n is 1.

Because the slope of line m is -1 and the slope of line n is 1, their slopes are opposite reciprocals. Therefore, line m is proven to be perpendicular to line n.

The Least You Need to Know

- Coordinate planes help you locate points on a plane.
- The distance formula can be used to find the distance between any two points, which is also the length of the segment joining those points.
- The midpoint formula can be used to find the midpoint of a segment or one of its endpoints.
- Slope can help you determine if lines on a coordinate plane are parallel or perpendicular.

PART 2

Reasoning and Proof

A teacher may share with you many problems and answers, but as the scope of mathematics grows, it is important to be able to answer the questions "how" and "why." This is where you need to understand proof. Aside from postulates, which are understood as fact, geometry consists of countless definitions and theorems that are used to prove statements. In this part, you learn how to use the elements from foundational geometry to write basic proofs.

CHAPTER 5

Inductive vs. Deductive Reasoning

Cause and effect is applied to your daily activities as you decide to clean your room, prepare a meal, or put your shoes on. Did you ever stop to think why you did what you did?

In mathematics, you use inductive and deductive reasoning to observe a visual or numerical pattern, recognize the pattern, and make assumptions. The ability to reason logically using inductive and deductive reasoning is important to the study of geometry. In this chapter, we explore relationships between two things, make assumptions about these relationships, and learn how to use inductive and deductive reasoning to explain why these assumptions are true.

In This Chapter

- Using inductive reasoning
- Writing the contrapositive, converse, and inverse
- Using a Euler diagram to interpret the validity of the statement
- Drawing conclusions based on conditional statements

Inductive Reasoning

Inductive reasoning is the process of drawing conclusions from experience. In mathematics, you use inductive reasoning to guess what might be true about a numerical pattern. For example, if you increase the number of 3s you multiply with by one more, you get a pattern:

$$3 = 3^1$$

$$3 \times 3 = 3^2$$

$$3 \times 3 \times 3 = 3^3$$

$$3 \times 3 \times 3 \times 3 = 3^4$$

One three is equivalent to 3^1, the product of two threes is 3^2, the product of three threes is 3^3, and so on. Can you predict how you would write the product of five threes in exponential form? If you said 3^5, you are on the right track!

Now, you might conclude that the exponent represents the number of 3s you are multiplying. Therefore, perhaps a number n to an exponent e, written n^e, will always equal the product of n taken e times.

Although you have observed a pattern, it does not mean it is true in all situations. If you recall from algebra, $x^0 = 1$ given x does not equal 0. Therefore, 3^0 is 1 and n^e will not always represent the product of n taken e times. This is known as a *counterexample*.

DEFINITION

An example that proves a statement false is a **counterexample.**

Can you think of a way to edit your conclusion so that 3^0 is not a counterexample? The following is how you can edit it to be true:

A number n raised to the power of a natural number e will always equal the product of this number taken e times.

Using inductive reasoning, you have simply assumed a pattern; you have not proven it. When you write a statement like the preceding based on observations such that the statement has not been proven, it is known as a *conjecture*.

For example, given the following pattern, make a conjecture about the sequence:

5, 9, 13, 17, 21, …

If you notice that you have to add 4 to get from one term to the next, you are correct. Based on your conclusion, you can say that to continue the pattern, you add 4 to the previous number in the sequence.

Now try using inductive reasoning and conjectures to explore the sum of the angles of a triangle. Measure each of the angles of the triangle in Figure 5.1 using a protractor. What do you notice about the sum of these angle measures? The sum should be 180°.

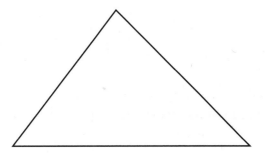

Figure 5.1: *Measuring the angles of a triangle.*

Next, draw two different triangles and measure the angles of each of these triangles. The sum of the angles of each triangle should also equal 180°. Therefore, what's your conjecture? Here's what you can use:

The sum of the angles of a triangle will always equal 180°.

Conditional Statements

When you are writing an assumption such as "The sum of the angles of a triangle will always equal 180°," how many ideas are in the statement? More than one! Had you written "The sum of the angles is 180°," you would have only one idea.

A sentence that combines two or more statements with words such as *and, or,* and *if ... then* is known as a *compound statement* or, for the purposes of geometry, a *conditional statement.*

Let's write the previous statement in if-then form:

If you are measuring the angles of a triangle, then the sum of these measures will result in 180°.

By writing the statement in the if-then form of a conditional statement, you are able to relate back to what you were observing.

The following are other examples of conditional statements:

> If the traffic light is green, then you must go.

> If you live in Springfield, then you live in Massachusetts.

> If three points are noncollinear, then the points are coplanar.

IT DOESN'T ADD UP

A conditional statement is a way to write an assumption; the statement isn't necessarily true. For example, Springfield may be in Massachusetts, but it could also turn out that the Springfield referred to is in another state.

Parts of a Conditional Statement

A conditional statement has two parts: a hypothesis and a conclusion. The *hypothesis,* also known as the *antecedent,* is the part after the "if"; the *conclusion,* also known as the *consequent,* is the part after the "then."

The lowercase letter p is commonly used to denote the hypothesis and q for the conclusion. So in the example "If the traffic light is green, then you must go," p = "the traffic light is green" and q = "you must go."

Now you have "If p, then q." What do you do with the "if ... then"? An arrow pointing to the right, read as "implies," replaces the "if ... then." Thus, conditional statements can be thought of as "$p{\rightarrow}q$," which reads "p implies q."

Verifying the Validity of Conditional Statements

Conditional statements can be true or false. If all cases that make the hypothesis true always make the conclusion true, then the conditional statement is true. To show that a conditional statement is false, you need one true hypothesis that proves the conclusion false—that is, a counterexample.

Let's rewrite the example "You live in New Jersey if you live in Newark" in if-then form:

> If you live in Newark, then you live in New Jersey.

But Newark is also the name of a city in Delaware; this is considered a counterexample. Therefore, the preceding conditional statement is false.

It is not always easy to process whether a conditional is true or false, so a helpful way to analyze a conditional statement is to model it using a *Euler diagram.* A Euler diagram is a way of picturing relationships between different groups of things.

The Euler diagram in Figure 5.2 shows a relationship between "if *p*, then *q*."

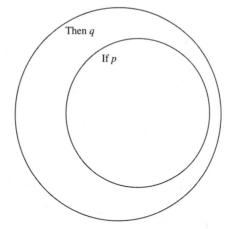

Figure 5.2: *The visual representation of "if p, then q."*

Let's analyze whether a statement is true or false using the Euler diagram to guide your thought process. In the Euler diagram in Figure 5.3, the set of rectangles is part of the set of quadrilaterals. Therefore, the conditional statement associated with this diagram is "If the shape is a rectangle, then it is a quadrilateral." Related to this statement, can you think of a rectangle that is not a quadrilateral?

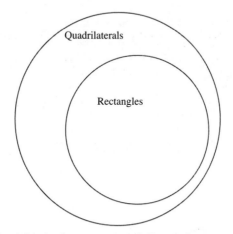

Figure 5.3: *The visual representation of "if rectangle, then quadrilateral."*

The answer is no, because all rectangles are four-sided figures, which means all rectangles are quadrilaterals. Therefore, the conditional statement is true.

What if the Euler diagram shows the set of quadrilaterals as a subset of rectangles, like in Figure 5.4? Now the statement would be "If the shape is a quadrilateral, then it is a rectangle." Can you think of a four-sided figure that is a quadrilateral but not classified as a rectangle?

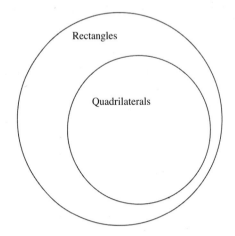

Figure 5.4: *The visual representation of "if quadrilateral, then rectangle."*

What about a trapezoid? Because a trapezoid is defined as a quadrilateral with one pair of parallel sides, it is not a rectangle. Thus, this conditional statement is false.

Let's look at one more conditional statement with a Euler diagram: "If a triangle has three acute angles, then it is an equilateral triangle." Think about triangles with three acute angles that are the same and three acute angles that are different—for example, a triangle with three angles of 60° and a triangle with angles of 61°, 59°, and 60°. Where would you place each of these examples? Figure 5.5 shows you.

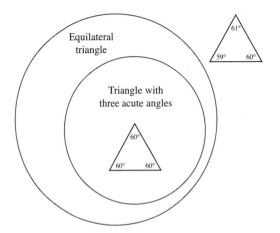

Figure 5.5: *A triangle with three acute angles does not imply that the triangle is an equilateral triangle.*

The triangle with three acute angles is a subset of the group of equilateral triangles. Because the triangle with three angles measuring 60° degrees satisfies the hypothesis and the conclusion, you would place it in the inner circle. The triangle with angles measuring 61°, 59°, and 60° does not satisfy the conclusion; therefore, you write it in the outer circle. Because you found a counterexample, the conditional statement is false.

Let's take a look at a few more examples. This time, the conditional statements are not in if-then form. Practice writing them in if-then form and then decide whether the statement is true or false. If it's false, provide a counterexample.

👆 **HELPFUL POINT**

When you write the statement in if-then form, you may need to clarify the hypothesis and conclusion by defining the subject of the statement.

The first statement is "A guitar player is a musician." For this, p is "A guitar player" and q is "is a musician." In if-then form, the statement would be written as "If a person is a guitar player, then the person is a musician."

In order to determine if this statement is true or false, can you think of a counterexample? Is a guitar player ever not a musician? The answer is no. Therefore, the statement is true.

Here's another statement: "The value of x^2 is always greater than x." In this case, p is "x is squared" and q is "the value of x^2 is always greater than x." In if-then form, the statement would be written as "If x is squared, then the value of x^2 will always be greater than x."

Because x can be any number, let $x = 2$. Thus, x^2 is equal to 4, $4 > 2$, and $x^2 > x$. However, if $x = 1$, then $x^2 = 1$, and 1 is not greater than 1. Therefore, $x = 1$ is a counterexample and the statement is false.

Related Conditionals

Related conditionals can be used to write clear definitions. For example, consider the term *radius*. Here's a definition for it: "Given a segment whose endpoints are the center of circle and a point on the circle, the distance of this segment is the length of the radius." Pretty hard to follow, isn't it?

If you write it in if-then form using related conditionals, however, it becomes clearer: "If a segment's endpoints are the center of a circle and a point on the circle, then the distance of the segment is the length of the radius." So radius is the length of a segment whose endpoints are the center of a circle and a point on the circle.

In this section, we show you how to use related conditionals to write definitions.

Converses and Biconditionals

A clear definition can be written when a conditional and its *converse* are both true. The converse of a conditional statement $p \rightarrow q$ is $q \rightarrow p$—that is, the hypothesis and conclusion are switched. For example, for the statement "If the measure of an angle is 180°, then it is a straight angle," the converse would be "If an angle is a straight angle, then its measure is 180°." In this situation, the conditional statement and its converse are both true!

But can the conditional and its converse not be the same? Yes! For example, given the conditional statement "If a number is a whole number, then it is an integer," the converse is "If the number is an integer, then it is a whole number." For the conditional statement, can you think of a whole number that is not an integer? The answer is no. Therefore, the conditional statement is true. For the converse, can you think if an integer that is not a whole number? The answer is yes! How about -5? A number such as -5 is an integer but not a whole number. So the conditional is true, while the converse is false.

When $p \rightarrow q$ is true and $q \rightarrow p$ is true, you can write a *biconditional statement*. The abbreviation for a biconditional statement is $p \leftrightarrow q$, which is read as "p if and only if q." For example, you can use the "if and only if" form to write, "An angle is a straight angle if and only if its measure is 180°." However, you can't write a biconditional for the conditional "If a number is a whole number, then it is an integer" because the converse is not true.

Negations

Before we define the other two related conditionals, we need to examine *negations*. The negation, written as $\sim p$ or $\sim q$, of a statement is the opposite of the statement. Here are some examples:

Statement: The dog is white.

Negation: The dog is not white.

Statement: It is not raining.

Negation: It is raining.

HELPFUL POINT

Negation allows you to understand what something is by thinking about what it is not.

Now we can define the next related conditional, the *inverse*. The inverse of a conditional statement is written by negating the hypothesis and conclusion. It is shown as $\sim p \rightarrow \sim q$. For example, the inverse of "If a number is a whole number, then it is an integer" is "If a number is not a whole number, then it is not an integer." Just because the conditional statement is true does not mean the inverse is true. For example, -5 is not a whole number; however, it is an integer. This would show the statement is false.

Lastly, the *contrapositive* of a conditional statement, written as ~q→~p, is the converse of the inverse. For example, the contrapositive of "If a number is a whole number, then it is an integer," is "If a number is not an integer, then it is not a whole number." The contrapositive is true because the number between two integers will never be a whole number. Unlike the converse and inverse, if a conditional statement is true, its contrapositive will always be true.

Drawing Conclusions Using Related Conditionals

Now that you know how conditionals relate to each other, you can use them to draw certain conclusions. For example, take a look at the conditional "If a person is a basketball player, then the person is an athlete."

With the following pieces of information, see what you can conclude about each person, if anything: Dan is a basketball player, Jose is an athlete, Mary is not a basketball player, and Rickey is not an athlete. Using a Euler diagram, you could write how these apply to the conditional statement, as shown in Figure 5.6. In this case, you should construct a Euler diagram in which the inner circle is the basketball player and the outer circle is the athlete.

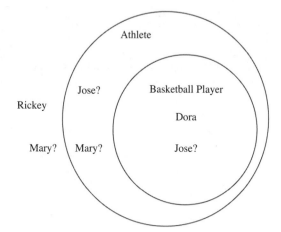

Figure 5.6: *A Euler diagram for the conditional statement "If a person is a basketball player, then the person is an athlete."*

As you can see in the figure, because Dan is a basketball player, he is in the inner circle. And because p→q, you can conclude that Dan is an athlete.

For Jose, you don't know if he is a basketball player or some other type of an athlete. Therefore, you can't draw a conclusion. Furthermore, the converse of the conditional "If a person is an athlete, then that person is a basketball player" can't be assumed true.

With Mary, you don't know if she is or isn't an athlete. Because you do not know where to place her, you can't draw a conclusion. Furthermore, the inverse of the conditional "If a person is not a basketball player, then the person is not an athlete" can't be assumed true.

Finally, because Rickey is not even an athlete, he is placed outside the outer circle. In this case, $\sim q \rightarrow \sim p$, so you can conclude Rickey is not a basketball player because the contrapositive of a true conditional—the contrapositive here being "If a person is not an athlete, then the person is not a basketball player." —will always be true.

Deductive Reasoning

Looking for a pattern does not guarantee that a statement is always true. It is necessary to prove a statement in order to assume that it is absolutely true for all cases. *Deductive reasoning* is the process of using facts, definitions, rules, properties, and the *laws of logic* to form a logical argument. This is different from inductive reasoning, which uses specific examples and patterns to form conjectures.

DEFINITION

The **laws of logic,** also known as De Morgan's laws, are used to make valid inferences in proofs and deductive arguments. The two laws are the Law of Detachment and the Law of Syllogism.

One of the laws of logic is the Law of Detachment, which is as follows:

If $p \rightarrow q$ is a true conditional and p is true, then q is true.

Referring back to the basketball player scenario in the previous section, let's use the Law of Detachment to make a valid conclusion. As you know, the true conditional statement is "If a person is a basketball player, then the person is an athlete." You know that Dan is a basketball player. Because p is also true, you can conclude that q is true; therefore, Dan is an athlete, too.

The second law of logic is the Law of Syllogism, which is as follows:

If $p \rightarrow q$ is true and $q \rightarrow r$ is true, then $p \rightarrow r$ is true.

To understand Law of Syllogism, write a new conditional statement that follows from this pair of true statements:

If it is cloudy outside, it will rain.

If it rains outside, you will need an umbrella.

Because both statements are true, you can combine *p*, "If it is cloudy outside," with *r*, "you will need an umbrella." Therefore, "If it is cloudy outside, you will need an umbrella."

You will extend deductive reasoning using properties of algebra to prove statements about segments, angles, and angle pairs in Chapter 6.

The Least You Need to Know

- In inductive reasoning, you are making an assumption. In deductive reasoning, you attempt to prove the assumption to be true.
- Using a Euler diagram can be helpful in visualizing conditional statements.
- If a conditional statement is true, its contrapositive is also true.

Creating Two-Column Proofs

An architect presents the ideas for building a home through his blueprint. A contractor turns a scale drawing into an actual home. Likewise, thus far, we have presented you with the basics of geometry. You are now ready to take the understanding of the basic concepts to the next level. As contractors of geometry, you will build understanding of these concepts by proving important conjectures in the context of a logical system.

In this chapter, you learn about two-column proofs and how to apply algebraic properties of equality, as well as definitions and postulates of geometry, in order to prove assumptions.

Developing a Proof

The purpose of a proof is to prove a statement using what is given along with properties to justify each step leading to the conclusion. While developing a proof can be a daunting task, planning the proof helps. So how do you get started?

The following are some strategies to brainstorm how you can go from what is given to what you are trying to prove:

1. Visualize what is true. Draw and label any diagrams based on what is provided in the given.

2. Represent the situation algebraically.

3. Work backward. Remember, you are not looking for the answer; you are looking to show how to get to the answer.

> **HELPFUL POINT**
>
> How does the second strategy work? Suppose you are given a right triangle with a second angle measuring 30°. You can prove that the measure of the third angle is 60° by writing the equation $x + 90° + 30° = 180°$.

A two-column proof is the most common way to lay out a proof. To fill one out, you start with the *given*, or what is true about the problem, followed by the statement you need to prove based on the given. The first column is where you write the statements, starting with the given and ending with what you are proving. The second column is where you will reason how you got from one step to the next using definitions, properties, postulates, and other theorems.

Algebraic Proofs

The process of deductive reasoning is used to solve an equation. Therefore, before we go through an example of a two-column algebraic proof, let's revisit the properties of equality to help you understand what you'll be doing when filling out proofs.

Addition Property of Equality: If $a = b$, then $a + c = b + c$.

For example, if $x - 5 = 10$, then $(x - 5) + 5 = (10) + 5$.

Subtraction Property of Equality: If $a = b$, then $a - c = b - c$.

For example, if $x + 5 = 10$, then $(x + 5) - 5 = (10) - 5$.

Multiplication Property of Equality: If $a = b$, then $ac = bc$.

For example, if $0.5x = 10$, then $2(0.5x) = 2(10)$.

Division Property of Equality: If $a = b$ and c do not equal 0, then $a \div c = b \div c$.

For example, if $5x = 10$, then $\dfrac{5x}{5} = \dfrac{10}{5}$.

Reflexive Axiom of Equality: $a = a$.

For example, 5 must equal 5.

Symmetric Axiom of Equality: If $a = b$, then $b = a$.

For example, if $2(5) = 10$, then $10 = 2(5)$.

Transitive Axiom of Equality: If $a = b$ and $b = c$, then $a = c$.

For example, if $2(4) = 8$ and $8 = 1(8)$, then $2(4) = 1(8)$.

Substitution Property of Equality: If $a = b$, then a can be replaced with b in an equation or expression.

For example, if $x + 2 = 8$ and $x = 6$, then $6 + 2 = 8$.

Distributive Property of Equality: $a(b + c) = ab + ac$.

For example, $2(1 + 6)$ must equal $2(1) + 2(6)$.

Now that you've reviewed the algebraic properties of equality, let's use these properties to reason how to solve each equation in a proof. If you recall, you should write the given information and any statements on the left and the reasons and proof on the right.

Given: $3x + 6 = 18$

Prove: $x = 4$

Statement	Reason
1. $3x + 6 = 18$	1. Given
2. $3x = 12$	2. Subtraction property of equality
3. $x = 4$	3. Division property of equality

To get from step 1 to step 2, you subtract 6 from both sides. Doing so is justified by the subtraction property of equality. Next, you divide both sides by 3. Thus, you apply the division property of equality and arrive at what you need to prove.

HELPFUL POINT

It's okay to reference the properties of equality if you get stuck! That's how you'll familiarize yourself with them.

Let's try another one. This time, you fill in the reasons.

Given: $\frac{1}{2}(6x - 12) = -12$

Prove: $x = -2$

Statement	Reason
1. $\frac{1}{2}(6x - 12) = -12$	1.
2. $3x - 6 = -12$	2.
3. $3x = -6$	3.
4. $x = -2$	4.

What do you do to get from the given to step 2? You distribute the $\frac{1}{2}$. Therefore, your reason for step 2 should be the distributive property of equality. To get from step 2 to step 3, you add 6 to both sides, which is justified by the addition property of equality. To get from step 3 to step 4, you divide both sides by 3. Thus, you apply the division property of equality.

Let's try one more where you have to fill in blanks in both columns. Don't worry, you can do it!

Given: $3x^2 + 18x - 42 = 3x(x + 4)$

Prove: $x = 7$

Statement	Reason
1. $3x^2 + 18x - 42 = 3x(x + 4)$	1.
2.	2. Distributive property of equality
3. $18x - 42 = 12x$	3.
4. $-42 = -6x$	4.
5.	5. Division property of equality
6. $x = 7$	6.

In step 1, you should write "Given" in the right column, as the information in step 1 is what's given to you. Next, think about what you distribute in step 1 to get to step 2. That's right, you multiply the information in parentheses by $3x$ on the right side, so it's $3x^2 + 18x - 42 = 3x^2 + 12$. To get from step 2 to step 3, you subtract $3x^2$ from both sides, which is justified by the subtraction property of equality. From step 3 to step 4, you subtract $18x$ from both sides, again applying the subtraction property of equality. To get to step 5, you use the division property of equality, which means you must divide both sides by -6. Therefore, $7 = x$, which you can rewrite as $x = 7$. Looking back at the properties, this is the symmetric property of equality. And those are all the steps!

Geometric Proofs

Writing geometric proofs is pretty similar to writing algebraic proofs—the only real difference is you're talking about angles and segments this time. Let's take a closer look.

Proofs Involving Segments

You should know that the properties of equality used earlier in the chapter are also true for segment lengths. You should also look back at Chapters 3 about segment addition for a refresher on concepts related to segments. For a proof involving segments, let's start with a two-step proof.

Given: $AB = XY$

Prove: $\overline{AB} \cong \overline{XY}$

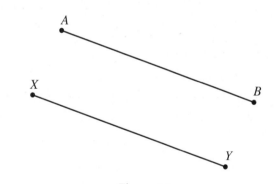

Figure 6.1

Statement	Reason
1. $AB = XY$	1. Given
2. $\overline{AB} \cong \overline{XY}$	2. Definition of congruent segments

The definition of congruent segments states that segments that have the same length are congruent. You are given the measure of two segments that are equal; therefore, by the definition of congruent segments, $\overline{AB} \cong \overline{XY}$.

Next, take a look at this slightly longer proof that requires the Segment Addition Postulate, as well as algebraic properties of equality.

Given: $AB = CD$

Prove: $AC = BD$

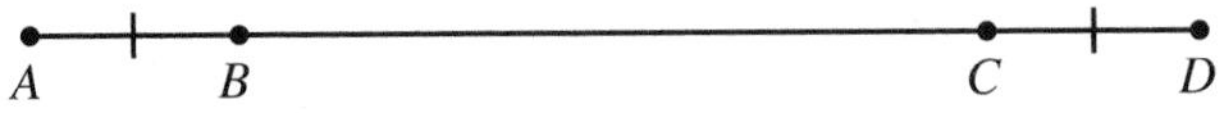

Figure 6.2

Statement	Reason
1. $AB = CD$	Given
2. $BC = BC$	Reflexive axiom of equality
3. $AB + BC = CD + BC$	Addition property of equality
4. $AB + BC = AC; BC + CD = BD$	Segment Addition Postulate
5. $AC = BD$	Substitution property of equality

To start, look at Figure 6.2 and see how you can get from the given of $AB = CD$ to $AC = BD$. As you can see in the figure, both $\overline{AC}$ and $\overline{BD}$ include the $\overline{BC}$. In step 2, you justify $BC = BC$ by the reflexive axiom of equality. Because it is the same value, what you add to one side of the given, you can add to the other; therefore, $AB + BC = CD + BC$ by the addition property of equality. Next, you know from Chapter 3 that the sum of the lengths of the shorter segments equals the length of the larger segment, meaning $AB + BC = AC$ and $BC + CD = BD$. This is the Segment Addition Postulate. Last, you replaced $AB + BC$ with AC and $CD + BC$ with BD, which is the substitution property of equality.

HELPFUL POINT

If you recall from Chapter 3, the Segment Addition Postulate states that if B is between A and C, then $AB + BC = AC$. So if $AB + BC = AC$, you know B is between A and C.

For this last segment proof, try to fill in the blanks.

Given: $AC = DF; BC = DE$

Prove: $\overline{AB} \cong \overline{EF}$

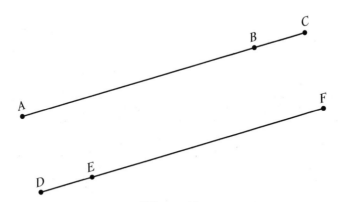

Figure 6.3

Statement	Reason
1. $AC = DF$	1. Given
2. ? $= AB + BC$ $DF = ? + ?$	2. Segment Addition Postulate
3. $AB + BC = DE + DF$	3.
4.	4. Given
5. $AB + BC = ? + DF$	5.
6. $AB = EF$	6.
7.	7. Definition of congruent segments

Because you use the Segment Addition Postulate in step 2, that means $AC = AB + BC$ and $DF = DE + EF$. You are then able to replace the segment lengths you are given in step 1 with what you conclude in step 2 to arrive at step 3, which is justified by the substitution property of equality. Step 4 has to be $BC = DE$ because you have not used this given yet. Now think, what can you do with this given and the equation in step 3? How about replacing the DE with BC? Therefore, step 5 should read $AB + BC = BC + DF$ by the substitution property of equality. Next, you subtract both sides by BC, making $AB = EF$ by the subtraction property of equality. Last, segments with the same lengths are congruent segments. Thus, you prove $\overline{AB} \cong \overline{EF}$ by the definition of congruent segments.

Proofs Involving Angles

You can also use the algebraic properties of equality to write geometry proofs involving angles. Recall the definition of congruent angles by looking at the following two-step proof for angles with the same measure.

Given: $m\angle 1 = m\angle 2$

Prove: $\angle 1 \cong \angle 2$

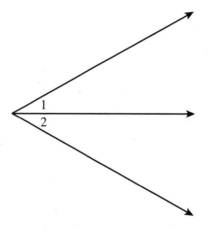

Figure 6.4

Statement	Reason
1. $m\angle 1 = m\angle 2$	1. Given
2. $\angle 1 \cong \angle 2$	2. Definition of congruent angles

The definition of congruent angles states that angles that have the same measure are congruent angles. Because you are given the measure of two angles that are equal, by the definition of angles segments, $\angle 1 \cong \angle 2$.

Next, let's review a slighter longer proof that requires the Angle Addition Postulate, as well as algebraic properties of equality.

Given: $m\angle XVZ = m\angle YVW$

Prove: $m\angle 1 = m\angle 3$

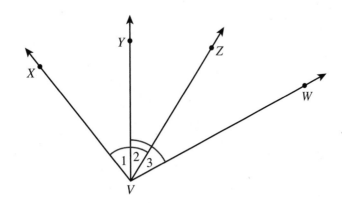

Figure 6.5

Statement	Reason
1. $m\angle XVZ = m\angle YVW$	1. Given
2. $m\angle XVZ = m\angle 1 + m\angle 2 + m\angle YVW = m\angle 2 + m\angle 3$	2. Angle Addition Postulate
3. $m\angle 1 + m\angle 2 = m\angle 2 + m\angle 3$	3. Substitution property of equality
4. $m\angle 2 = m\angle 2$	4. Reflexive axiom of equality
5. $m\angle 1 = m\angle 3$	5. Subtraction property of equality

If you recall angle addition from Chapter 3, one angle can be divided into two smaller angles, and the sum of the smaller angles equals the measure of the larger angle. The postulate that states this is the Angle Addition Postulate. Therefore, $m\angle XVZ$ equals the sum of $m\angle 1$ and $m\angle 2$. Furthermore, $m\angle YVW$ equals the sum of $m\angle 2$ and $m\angle 3$. Next, in step 3, you replaced the terms in step 1 with what you know from step 2. Finally, since by the reflexive axiom of equality the measure of an angle equals itself, you are able to subtract $m\angle 2$ from both sides and conclude $m\angle 1 = m\angle 3$.

Now recall the definition of a linear pair of angles. Linear pairs of angles are both adjacent and supplementary, which is known as the *Linear Pair Postulate*. Let's look at a proof that uses the Linear Pair Postulate.

 **DEFINITION**

The **Linear Pair Postulate** states that if two angles form a pair, they are supplementary.

Given: $m\angle 1 + m\angle 3 = 180°$

Prove: $m\angle 2 = m\angle 3$

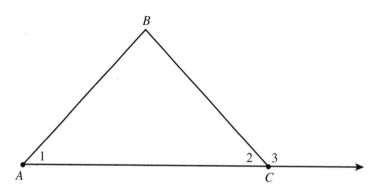

Figure 6.6

Statement	Reason
1. $m\angle 1 + m\angle 3 = 180°$	1. Given
2. $m\angle 2 + m\angle 3 = 180°$	2. Linear Pair Postulate
3. $m\angle 1 + m\angle 3 = m\angle 2 + m\angle 3$	3. Substitution property of equality
4. $m\angle 3 = m\angle 3$	4. Reflexive axiom of equality
5. $m\angle 1 = m\angle 2$	5. Subtraction property of equality

In the first step, as usual, you state the given. In the second step, you use the Linear Pair Postulate to figure that because the angles are adjacent, they are supplementary. In step 3, you substitute 180° with its equivalence from step 2. By noting that $m\angle 3 = m\angle 3$ using the reflexive axiom of equality in step 4, you are able to use the subtraction property of equality in step 3 to get to step 5.

Last, let's look at a proof that uses the definition of supplementary angles. If you recall, two angles are supplementary when the sum of their angle measures is 180°.

Given: $\angle 1$ and $\angle 2$ are supplements; $\angle 3$ and $\angle 4$ are supplements; $\angle 1 \cong \angle 4$

Prove: $\angle 2 \cong \angle 3$

Figure 6.7

Statement	Reason
1. $\angle 1$ and $\angle 2$ are supplements; $\angle 3$ and $\angle 4$ are supplements	1. Given
2. $m\angle 1 + m\angle 2 = 180$; $m\angle 3 + m\angle 4 = 180$	2. Definition of supplementary angles
3. $m\angle 1 + m\angle 2 = m\angle 3 + m\angle 4$	3. Substitution property of equality
4. $\angle 1 \cong \angle 4$	4. Given
5. $m\angle 1 = m\angle 4$	5. Definition of congruent angles
6. $m\angle 1 + m\angle 2 = m\angle 3 + m\angle 1$	6. Substitution property of equality
7. $m\angle 1 = m\angle 1$	7. Reflexive axiom of equality
8. $m\angle 2 = m\angle 3$	8. Subtraction property of equality
9. $\angle 2 \cong \angle 3$	9. Definition of congruent angles

By using the definition of supplementary angles, you are able to set the given supplements to 180°. Because both equations equal 180°, you can set the sums of the measures of the supplements equal to one another by the replacing 180° with $m\angle 3 + m\angle 4$. This is justified by the substitution property of equality. Next, you know that $\angle 1$ and $\angle 4$ are congruent, so you can use the definition of congruent angles to set $m\angle 1$ equal to $m\angle 4$. In step 6, you are substituting the 180° from step 2 and replacing the $m\angle 4$ with $m\angle 1$ using step 5. Because the $m\angle 1$ is equal to itself by the reflexive axiom of equality, you can subtract $m\angle 1$ from both sides of step 6and arrive at step 8. Thereafter, you simply use the definition of congruent angles.

HELPFUL POINT

Throughout the remainder of this book, you will continue to prove important conjectures in the context of a proof. Remember, practice makes perfect! You can take a blank piece of paper and try writing the proofs from this chapter without looking at the filled-in proofs.

The Least You Need to Know

- Writing a two-column proof is a formal way of organizing your statements to show a statement is true.
- You can use algebraic properties of equality to justify statements in geometric proofs.
- You can prove statements about segments and angles with both the algebraic properties of equality and the Segment Addition or Angle Addition Postulate.

Parallel and Perpendicular Lines

In Chapter 4, you explored the slopes of parallel and perpendicular lines, and in Chapter 6, you learned how to develop a proof. In this chapter, we take these a step further by investigating the angles that are formed as a line intersects parallel lines, as well as angle relationships of perpendicular lines. We then use these angle pair relationships to prove congruent angles.

Parallel Lines in Relation to Angles

You explored coplanar lines in Chapter 1 and parallel lines on a coordinate plane in Chapter 4, so here let's talk about how transversals and angle pair relationships are related to parallel lines. A *transversal* is a line that intersects two or more coplanar lines at different points. When the coplanar lines cut by a transversal are parallel, the angles formed have special relationships. For example, in Figure 7.1, line *m* is parallel to line *n*, which is written as $m \parallel n$. Line *t* represents the transversal.

In This Chapter

- Understanding angles formed by parallel lines and a transversal
- Using angle relationships to prove lines are parallel
- Learning about theorems related to perpendicular lines
- Using theorems to prove lines are perpendicular

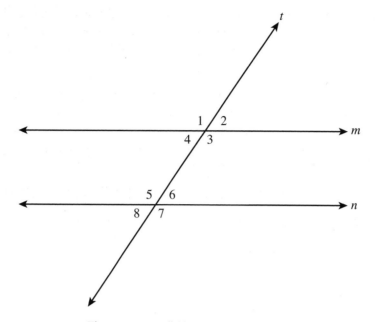

Figure 7.1: *Parallel lines cut by a transversal.*

Going by the angle numbers in Figure 7.1, the following are different angle pair relationships related to parallel lines:

Corresponding angles are the angle pairs on the same side of the transversal and in the same position relative to each of the parallel lines. In the figure, the corresponding angles are 1 and 5, 2 and 6, 3 and 7, and 4 and 8. The Corresponding Angles Postulate states that if two parallel lines are cut by a transversal, then the pairs of corresponding angles are congruent.

> ☞ **HELPFUL POINT**
>
> If the lines cut by the transversal are not parallel, the angle pair relationship is the same; however, the angles are not congruent.

Alternate interior angles are the congruent angle pairs inside the parallel lines and on opposite sides of the transversal. In the figure, the alternate interior angles are 4 and 6 and 3 and 5. The Alternate Interior Angles Theorem states that if two parallel lines are cut by a transversal, then the pairs of alternate interior angles are congruent.

Alternate exterior angles are the congruent angle pairs outside the parallel lines and on opposite sides of the transversal. In the figure, the alternate exterior angles are 1 and 7, and 2 and 8. The Alternate Exterior Angles Theorem states that if two parallel lines are cut by a transversal, then the pairs of alternate exterior angles are congruent.

Same-side interior angles, also known as *consecutive interior angles,* are the angle pairs inside the parallel lines on the same side of the transversal. In the figure, the same-side interior angles are 3 and 5, and 4 and 6. The Same-Side Interior Angles Theorem states that if two parallel lines are cut by a transversal, then the pairs of same-side interior angles are supplementary.

Using Parallel Angle Pair Relationships to Solve Problems

Let's apply these relationships to the angles in Figure 7.2. Take a look at the figure and find one pair of alternate interior angles, one pair of same-side interior angles, and one pair of corresponding angles.

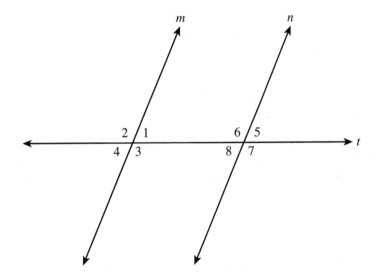

Figure 7.2: *Identifying angle pairs.*

In the figure, ∠ 1 and ∠ 8 are alternate interior angles because these angles are on opposing sides of the transversal and within the parallel lines. ∠ 1 and ∠ 6 are same-side interior angles because these angles are on the same side of the transversal and within the parallel lines. Finally, ∠ 2 and ∠ 6 are corresponding angles because these angles are on the same side of the transversal in the same position as it relates to each of the parallel lines.

Another thing you can use angle pair relationships for is to determine missing angle measures. For example, using Figure 7.2 again, determine the missing angle measures if $m\angle 1 = 60°$.

Given $m\angle 1 = 60°$, $m\angle 5 = 60°$ because $\angle 5$ is a corresponding angle to $\angle 1$. $m\angle 4$ must equal $m\angle 1$, and $m\angle 8$ must equal $m\angle 5$ based on the *Vertical Angles Congruence Theorem*. Therefore, $m\angle 4 = 60°$ and $m\angle 8 = 60°$.

DEFINITION

The **Vertical Angles Congruence Theorem** states that if two angles are vertical to one another, then they are congruent.

Next, recall the Linear Pair Postulate from Chapter 6:

$$m\angle 1 + m\angle 2 = 180°$$

$$60° + m\angle 2 = 180°$$

$$m\angle 2 = 120°$$

By correspondence, $m\angle 2$ would also equal 120°. By the Vertical Angles Congruence Theorem, $m\angle 3$ and $m\angle 7$ would also equal 120°. And those are all the angles in the figure!

Now that you have a handle on angle pair relationships, let's try an example involving algebra.

In Figure 7.3, determine the value of x.

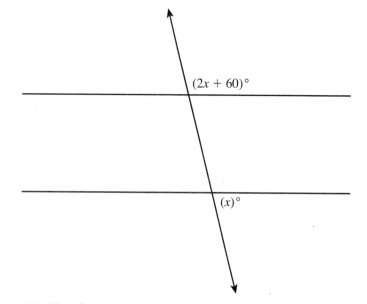

Figure 7.3: *What relationship do the angles in this figure have that can help you solve for x?*

The angle corresponding to the angle labeled $2x + 60$ forms a linear pair with the angle labeled x. Therefore, the corresponding angle must be $180 - x$. From here, you can write an equation and solve:

$$2x + 60 = 180 - x$$

$$3x = 120$$

$$x = 40$$

You can then plug in the answer to verify it:

$$2(40) + 60 = 180 - 40$$

$$80 + 60 = 160$$

$$160 = 160$$

Let's do one more together! In Figure 7.4, determine the value of the x.

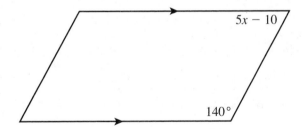

Figure 7.4: *Find out the relationship of the angles in this figure to solve for x.*

The arrows on the top and bottom of this figure indicate the segments are parallel to one another. Extend all of the segments to create lines. Do you see the angle pair relationship? The angles labeled $5x - 10$ and $140°$ are same-side interior angles. Now you can write and solve the equation for x:

$$5x - 10 + 140 = 180$$

$$5x + 130 = 180$$

$$5x = 50$$

$$x = 10$$

You can plug in the answer to verify it:

$$5(10) - 10 + 140 = 180$$

$$50 - 10 + 140 = 180$$

$$40 + 140 = 180$$

$$180 = 180$$

Proving Lines Are Parallel

You can use angle pair relationships to prove that lines are parallel. In these proofs, you'll be dealing with converses. If you recall from Chapter 5, the converse of a conditional statement is when you switch the hypothesis and conclusion. Remember that the converse of a true conditional statement does not need to be true. The theorems you learned from earlier on in this chapter all have true converses:

- **Corresponding Angles Converse:** If two lines are cut by a transversal so the corresponding angles are congruent, then the lines are parallel.

- **Alternate Interior Angles Converse:** If two lines are cut by a transversal so the alternate interior angles are congruent, then the lines are parallel.

- **Alternate Exterior Angles Converse:** If two lines are cut by a transversal so the alternate exterior angles are congruent, then the lines are parallel.

- **Same-Side Interior Angles Converse:** If two lines are cut by a transversal so the consecutive interior angles are supplementary, then the lines are parallel.

In this first example, let's prove the Corresponding Angles Converse to see how it works.

Given: $\angle 6 \cong \angle 7$

Prove: $m \| n$

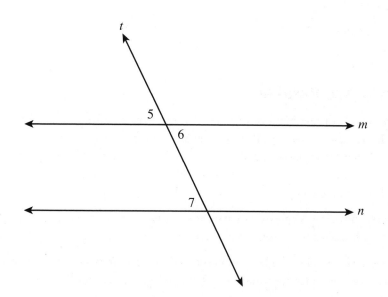

Figure 7.5: *Two lines intersected by a single line.*

Statement	Reason
1. $\angle 6 \cong \angle 7$	1. Given
2. $\angle 5 \cong \angle 6$	2. Vertical Angles Congruence Theorem
3. $\angle 5 \cong \angle 7$	3. Transitive property of congruence
4. $m \| n$	4. Corresponding Angles Converse

You are given that a pair of angles on the opposite sides of a transversal and inside the parallel lines are congruent. You can conclude $\angle 5 \cong \angle 6$ by the Vertical Angles Congruence Theorem. Thereafter, you apply the algebraic property of transitivity to write $\angle 5 \cong \angle 7$. Finally, you can state that the lines are parallel because two lines are parallel if there is a pair of corresponding angles, known as the Corresponding Angles Converse.

👉 **HELPFUL POINT**

When you are proving that measures of angles are equal, you use axioms of equality. When you are proving angles congruent, you use properties of congruence.

This next proof applies the Alternate Interior Angles Converse to prove two lines are parallel.

Given: $s \| t$ and $\angle 2 \cong \angle 1$

Prove: $m \| n$

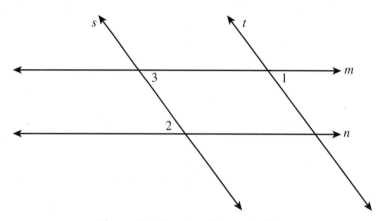

Figure 7.6: *Two pairs of intersecting lines.*

Statement	Reason
1. $s \| t$	1. Given
2. $\angle 1 \cong \angle 3$	2. Corresponding Angles Postulate
3. $\angle 2 \cong \angle 1$	3. Given
4. $\angle 2 \cong \angle 3$	4. Transitive property of congruence
5. $m \| n$	5. Alternate Interior Angles Converse

You know $s \| t$. By the Corresponding Angles Postulate, you can conclude $\angle 1$ is congruent to $\angle 3$. You also know that $\angle 2$ is congruent to $\angle 1$. The transitive property of equality states "If $a = b$ and $b = c$, then $a = c$"; the same holds true for congruence. Therefore, using steps 2 and 3, you can conclude that $\angle 2$ is congruent to $\angle 3$ by the transitive property of congruence. Finally, because $\angle 2$ and $\angle 3$ are alternate interior angles, you can use the Alternate Interior Angles Converse to conclude $m \| n$.

Perpendicular Lines in Relation to Angles

Recall the definition of a right angle from Chapter 2. A right angle is formed when two lines, segments, or rays intersect at a 90° angle.

For example, in Figure 7.7, $\overleftrightarrow{UZ}$ intersects $\overleftrightarrow{XW}$ at a 90° angle; therefore, these lines are also perpendicular to one another.

HELPFUL POINT

The symbol used to denote that two lines are perpendicular is $\perp$. Notice how it appears that one line intersects the other at a 90° angle.

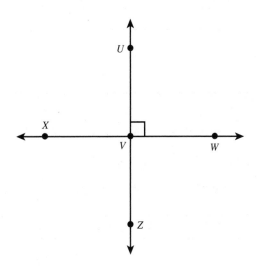

Figure 7.7: *An example of perpendicular lines.*

There are some important theorems related to perpendicular lines. First, the *Linear Pair of Congruent Angles Theorem* says, "If two lines intersect to form a linear of congruent angles, then the lines are perpendicular."

Second, the $\perp$ *Form Four Right Angles Theorem* says, "If two lines are perpendicular to each other, then they intersect to form four right angles." This is due to the supplement of a right angle being a right angle. Figure 7.8 shows an example of this theorem.

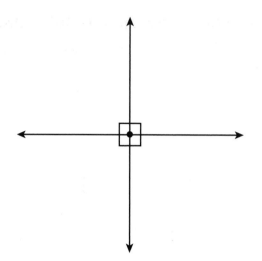

Figure 7.8: *An example of the ⊥ Form Four Right Angles Theorem.*

Last, the *Perpendicular and Complementary Theorem* says, "If two sides of two adjacent angles are perpendicular, then the angles are complementary." For example, in Figure 7.9, ∠ 1 is adjacent to ∠ 2 and $\overrightarrow{YX}$ is perpendicular to $\overrightarrow{YZ}$. Because the angle formed measures 90°, the adjacent angles are complementary.

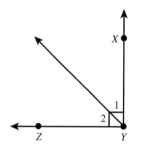

Figure 7.9: *An example of the Perpendicular and Complementary Theorem.*

Using Perpendicular Line Theorems to Solve Problems

Now that you have a handle on the angle relationships of perpendicular lines, let's try some examples involving algebra.

In Figure 7.10, determine the value of x, given $m \perp t$ and $m \parallel n$.

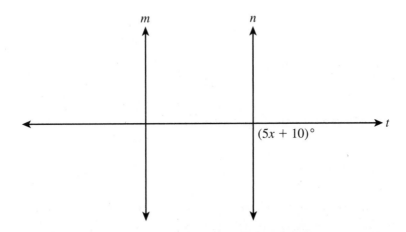

Figure 7.10: *Finding the value of x, given a line m is perpendicular to t and parallel to line n.*

Because m is parallel to n, n must also be perpendicular to t. Now you can write an equation and solve:

$$5x + 10 = 90$$

$$5x = 80$$

$$x = 16$$

You can now substitute the answer in to check your work:

$$5(16) + 10 = 90$$

Proving Lines Are Perpendicular

You can use the definition of and theorems related to perpendicular lines to prove that lines are perpendicular. The following is a proof that shows you how.

Given: $\angle 1$ and $\angle 2$ are complementary angles

Prove: $\overleftrightarrow{AB} \perp \overleftrightarrow{BC}$

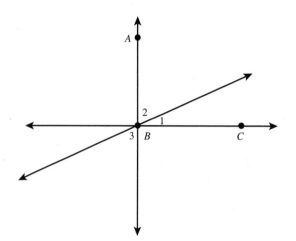

Figure 7.11: *Three lines that intersect at a single point.*

Statement	Reason
1. $\angle 1$ and $\angle 2$ are complementary angles	1. Given
2. $m\angle 1 + m\angle 2 = 90°$	2. Definition of complementary angles
3. $m\angle 1 + m\angle 2 = m\angle ABC$	3. Angle Addition Postulate
4. $m\angle ABC = 90°$	4. Substitution property of equality
5. $\overleftrightarrow{AB} \perp \overleftrightarrow{BC}$	5) Definition of perpendicular lines

Because $\angle 1$ and $\angle 2$ are complementary, you can conclude that the measures of these angles sum to 90° by the definition of complementary angles. By the Angle Addition Postulate, you can also conclude that the sum of $m\angle 1$ and $m\angle 2$ is $m\angle ABC$. Using steps 2 and 3, you can conclude that $m\angle ABC$ is 90° by the substitution property of equality. Finally, you know that two lines are perpendicular if they intersect at a 90° angle; therefore, $\overleftrightarrow{AB} \perp \overleftrightarrow{BC}$ by the definition of perpendicular lines.

Let's try one more proof.

Given: $m \perp n$, $\angle 1$, and $\angle 2$ are complementary

Prove: $\angle 3 \cong \angle 4$

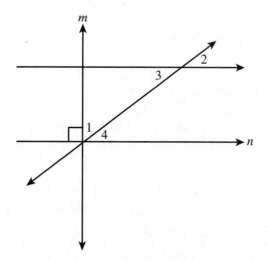

Figure 7.12: *Angles formed by intersecting lines.*

Statement	Reason
1. $m \perp n$, $\angle 1$ and $\angle 2$ are complementary	1. Given
2. $\angle 1$ and $\angle 4$ are complementary	2. Perpendicular Complementary Theorem
3. $m\angle 1 + m\angle 2 = 90°$ $m\angle 1 + m\angle 4 = 90°$	3. Definition of complementary angles
4. $m\angle 3 = m\angle 2$	4. Definition of vertical angles
5. $m\angle 1 + m\angle 3 = 90°$	5. Substitution property of equality
6. $m\angle 1 + m\angle 3 = m\angle 1 + m\angle 4$	6. Substitution property of equality
7. $m\angle 1 = m\angle 1$	7. Reflexive axiom of equality
8. $m\angle 3 = m\angle 4$	8. Subtraction property of equality
9. $\angle 3 \cong \angle 4$	9. Definition of congruent angles

Because two sides of two adjacent angles—namely, $\angle 1$ and $\angle 4$—are perpendicular based on the given, in step 2, you can conclude that $\angle 1$ and $\angle 4$ are complementary. When two angles are complementary, the sum of their measures is 90°, justifying step 3. Because $m\angle 3$ equals $m\angle 2$ by the definition of vertical angles, in step 5, you can substitute $m\angle 2$ with $m\angle 3$. Thereafter, you can use step 3 and step 5 to substitute for 90°. Because $m\angle 1$ is equivalent to itself by the reflexive axiom of equality, you can use the subtraction property of equality in step 6 to attain step 8. Finally, because $m\angle 3$ is equivalent to $m\angle 4$, $\angle 3 \cong \angle 4$.

The Least You Need to Know

- Corresponding, alternate interior, alternate exterior, or same-side interior angles formed by two parallel lines cut by a transversal are congruent.
- The converse of the corresponding angles, alternate interior angles, alternate exterior angles, and same-side interior angles theorems are all true.
- Theorems related to perpendicular lines can be used to justify congruent angle measures.

PART 3

Triangles

There are many shapes in the world, but in your study of geometry, you get to see just how complex some of these shapes can be. We start off your study of shapes with triangles. Not only are there several types of triangles, but they are also the catalyst for many geometric proofs. In this part, you learn about the relationships that exist between the side lengths and angle measures of triangles, which will help to broaden the basic proofs that you did in Part 2.

CHAPTER 8

Introducing Triangles

You have been learning about triangles since you were a child. However, geometry is about knowing more than that a triangle has three sides and three angles—you also need to know how it's classified and the relationships between those sides and angles. In this chapter, we give you lots of new things you need to know about triangles, including how they're classified, the relationship of segments to triangles, and triangle inequality theorems.

In This Chapter

- Classifying triangles
- Using concurrent lines and segments to find missing sides or angles of triangles
- Inequality theorems for triangles

Triangle Classifications

Just as you have a first name and a last name, triangles have two classifications related to their sides and their angles.

The classifications for sides can be thought of as the first name. When it comes to sides, triangles are *equilateral, isosceles,* or *scalene.*

An equilateral triangle has three congruent sides, an isosceles triangle has two congruent sides, and a scalene triangle has no congruent sides.

A triangle's "last name" comes from the classification of its angles as *acute, obtuse, right,* or *equiangular.* An acute triangle has three acute angles, an obtuse triangle has one obtuse angle, a right triangle has one right angle, and an equiangular triangle has three angles that are congruent.

There are a few important things to remember when it comes to the angles and sides of a triangle. If you recall, the sum of the interior angles of a triangle is 180°. So an equiangular triangle has three angles that each measure 60° because the three angles must be congruent and add up to 180°. Furthermore, all sides must also be congruent because the angles are congruent. This means that all equiangular triangles are also equilateral.

When it comes to obtuse triangles, a triangle is obtuse when it has one obtuse angle (an angle greater than 90°). Because triangles can't have more than one angle greater than 90°, it is impossible for a triangle to have more than one obtuse angle. The sides of obtuse triangles are classified as either isosceles or scalene because only two sides, at most, are congruent.

As you can see, the number of congruent sides and angles are directly related—a triangle with three congruent angles has three congruent sides, a triangle with two congruent angles has two congruent sides, and a triangle with no congruent angles has no congruent sides.

For example, if the angles of a triangle measure 45°, 45°, and 90°, you can see it has two congruent angles. Because it has two congruent angles, you know it also has two congruent sides. A triangle with two congruent sides is classified as isosceles. Now looking at the classification by angles, the triangle has a 90° angle, which is a right angle. Therefore, a triangle with angles measuring 45°, 45°, and 90° is an isosceles right triangle.

Now let's try an example where you find the missing angle measures of a triangle in Figure 8.1 and then classify it based on the answers.

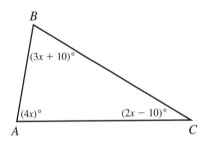

Figure 8.1: *Algebraic expressions indicate the measure of each angle in the triangle.*

Because the sum of the interior angles is 180°, the following equation can be written as follows:

$$(4x) + (3x + 10) + (2x - 10) = 180$$

Solve the equation:

$$(4x + 3x + 2x) + (10 - 10) = 180$$

$$9x = 180$$

$$\frac{9x}{9} = \frac{180}{9}$$

$$x = 20$$

The value of x can be substituted into the expression for each of the angles to find the angle measures:

$$m\angle A = 4x$$
$$m\angle A = 4(20)$$
$$m\angle A = 80°$$

$$m\angle B = 3x + 10$$
$$m\angle B = 3(20) + 10$$
$$m\angle B = 60 + 10$$
$$m\angle B = 70°$$

$$m\angle C = 2x - 10$$
$$m\angle C = 2(20) - 10$$
$$m\angle C = 40 - 10$$
$$m\angle C = 30°$$

 **HELPFUL POINT**

When finding angles in a triangle, you can always check your work by adding the three angles together to make sure the sum is 180°.

As you can see, none of the angles are congruent, meaning none of the sides are congruent. This means the triangle can be classified as scalene. Because the angles aren't congruent and are all under 90° or acute, the triangle can be classified by angles as an acute triangle. So the triangle in Figure 8.1 is a scalene acute triangle.

Concurrent Lines and Segments Related to Triangles

There are several *concurrent lines* and segments associated with triangles. A triangle has three of each type of these, with a point of intersection known generally as the *point of concurrency*. The following sections take you through each type and how you can use them to solve problems.

Altitudes

An altitude of a triangle is a perpendicular segment from a *vertex* to its opposite side; you can see an example of one in Figure 8.2. An altitude can be drawn from any vertex; therefore, every triangle has three altitudes. These altitudes are also referred to as the height of the triangle.

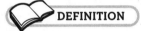

DEFINITION

Concurrent lines are three or more lines that intersect at the same point, called the **point of concurrency.** A **vertex** is the point at the corner or intersection of a geometric shape.

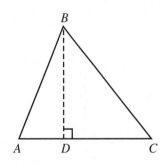

Figure 8.2: $\overline{BD}$ *is an altitude.*

When the three altitudes of a triangle intersect, their point of intersection is called an *orthocenter*. An orthocenter is a point of concurrency that can occur inside, on, or outside the triangle, depending on the types of angles in it. For example, in Figure 8.3, the altitudes of the right angle and two acute angles create an orthocenter on $\triangle ABC$ at point A. If all of the angles had been acute, the orthocenter would be inside the triangle; if one of the angles had been obtuse, the orthocenter would be outside the triangle.

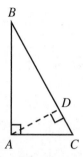

Figure 8.3: *For this right triangle, the three altitudes intersect at point A.*

Let's look at an example. Using the triangle in Figure 8.4, find the coordinates of the orthocenter.

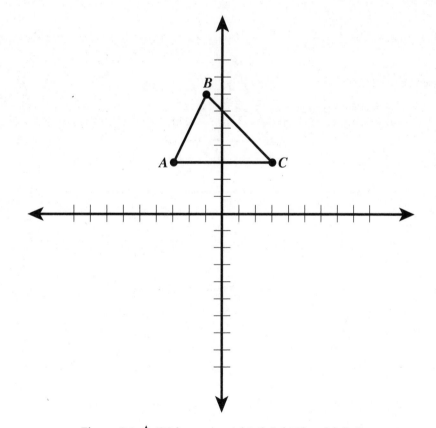

Figure 8.4: $\triangle ABC$ *has vertices A (-3, 3), B (-1, 7), and C (3, 3).*

To find the orthocenter, you need to find the coordinates of the intersection of two of the altitudes. First, identify the altitude that originates at vertex *B*. The altitude can be drawn directly down to intersect $\overline{AC}$ at (-1, 3). The equation of this line is $x = -1$.

Now look at the altitude that originates from vertex *A*. It is not as easy to identify a line perpendicular to $\overline{BC}$. Because you know that perpendicular lines have slopes that are opposite reciprocals, you can use the slope of $\overline{BC}$ to help you find the slope of the altitude that originates from vertex *A*. To find the slope of $\overline{BC}$, you can count the rise over run from the graph. To move from point *B* to point *C*, you need to move 4 spaces down (-4) and 4 spaces right (4). Because the rise is -4 and the run is 4, the slope of $\overline{BC}$ is $\frac{-4}{4} = -1$.

Because the slope of $\overline{BC}$ is -1, the slope of the altitude that originates at vertex *A* must be 1. Starting at vertex *A* and counting a rise of 1 and a run of 1, the altitudes will intersect at (-1, 5), as shown in Figure 8.5.

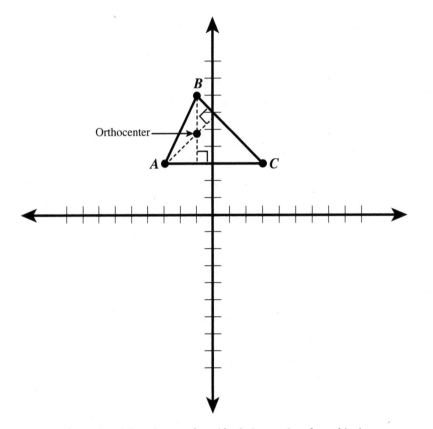

Figure 8.5: *The orthocenter formed by the intersection of two altitudes.*

Medians

The median of a triangle is a line segment joining a vertex to the midpoint of its opposite side, as shown in Figure 8.6. Because a median can be drawn from any vertex, every triangle has three medians.

HELPFUL POINT

Unlike altitudes, medians don't form a right angle with the side they intersect. Instead, each divides the triangle into two smaller triangles of equal area.

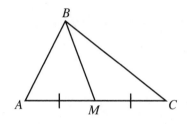

Figure 8.6: $\overline{BM}$ *is a median of* $\triangle ABC$.

The point of concurrency of the three medians of a triangle is called the *centroid*. The centroid is always inside the triangle and is two thirds of the distance from each vertex to the midpoint of the opposite side. For example, in Figure 8.7, *P* is the centroid of $\triangle ABC$.

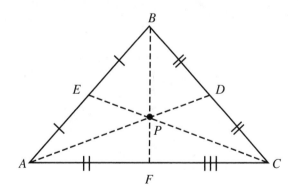

Figure 8.7: *The three medians of* $\triangle ABC$ *intersect at point P.*

The following equations are true based on $\triangle ABC$: $AP = \dfrac{2}{3} AD$, $BP = \dfrac{2}{3} BF$, and $CP = \dfrac{2}{3} CE$.

Now that you know the relationships formed by a centroid, let's look at an example. In the previous figure, you know *P* is the centroid of $\triangle ABC$. If *BP* = 10, find *PF* and *BF*.

Because *P* is the centroid, $BP = \dfrac{2}{3} BF$. Therefore, substitute the *BP* = 10 into the equation:

$$10 = \frac{2}{3} BF$$

$$\frac{3}{2}(10) = \frac{3}{2}\left(\frac{2}{3} BF\right)$$

$$15 = BF$$

You can then find the length of *PF* by subtracting *BP* from *BF*:

$$PF = 15 - 10$$

$$PF = 5$$

Angle Bisectors

An angle bisector in a triangle is a segment that is drawn from a vertex and cuts the vertex angle in half, as shown in Figure 8.8. Just like altitudes and medians, every triangle has three angle bisectors because they originate from each vertex.

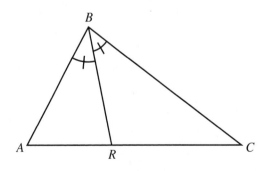

Figure 8.8: $\overline{BR}$ *is an angle bisector.*

The point of concurrency of the three angle bisectors of a triangle is called the *incenter.* The incenter is always inside the triangle and is equidistant from each of the three sides of a triangle. In Figure 8.9, *P* is the incenter of $\triangle ABC$, meaning *PD* = *PE* = *PF*.

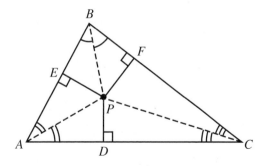

Figure 8.9: *Three angle bisectors intersect at point P.*

Let's look at an example using the previous figure. If $m\angle EBF = 80°$, find $m\angle BPE$.

Because $\overline{BP}$ is an angle bisector, $m\angle EPB = 40°$. Because the sum of the interior angles of ΔEBP is 180°, $m\angle EBP + m\angle BPE + m\angle PEB = 180°$. Substitute the angle measures you have into the equation:

$$40° + m\angle BPE + 90° = 180°$$

$$130° + m\angle BPE = 180°$$

$$130° - 130° + m\angle BPE = 180° - 130°$$

$$m\angle BPE = 50°$$

Perpendicular Bisectors

A perpendicular bisector of a triangle is a line that passes through the midpoint of a side and is perpendicular to that given side, as shown in Figure 8.10. Every triangle has three perpendicular bisectors.

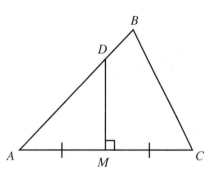

Figure 8.10: $\overline{DM}$ *is a perpendicular bisector.*

The point of concurrency of the three perpendicular bisectors of a triangle is called the *circumcenter*. The circumcenter can be inside, on, or outside the triangle and is equidistant from the vertices of a triangle. In Figure 8.11, *P* is the circumcenter of △*ABC*.

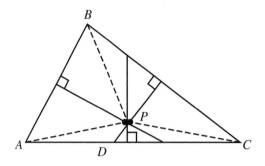

Figure 8.11: *Three perpendicular bisectors intersect at point P.*

Based on △*ABC*, *PA* = *PB* = *PC*.

Midsegments

A midsegment of a triangle connects the midpoints of two sides of a triangle. When drawn, the midsegment is parallel to the third side of the triangle, as you can see in Figure 8.12.

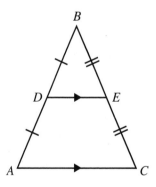

Figure 8.12: *Midsegment* $\overline{DE}$ *intersects two sides of the triangle.*

HELPFUL HINT

The arrows on $\overline{DE}$ and $\overline{AC}$ show they are parallel.

Because $\overline{DE}$ is a midsegment of $\triangle ABC$, D is the midpoint of $\overline{AB}$ and E is the midpoint of $\overline{BC}$. According to the Triangle Midsegment Theorem, $\overline{DE}$ and $\overline{AC}$ are parallel and $\overline{AC}$ is half the length of $\overline{AC}$.

But what is this theorem? The Triangle Midsegment Theorem states that a midsegment of a triangle is parallel to one side of the triangle, and its length is half the length of that side. With this theorem in mind, let's take a look at an example.

Using the previous figure, if $DE = 7$ inches, find AC.

According to the Triangle Midsegment Theorem, $DE = \dfrac{1}{2}AC$. Therefore, substitute the number into the equation and solve:

$$DE = \frac{1}{2}AC$$

$$7 = \frac{1}{2}AC$$

$$2(7) = 2\left(\frac{1}{2}AC\right)$$

$$14 = AC$$

Proportional Segments

You can draw lines in triangles that produce proportional segments. If you remember, *proportional* means "ratios that are equal" and a *ratio* is "a comparison between two quantities," so proportional segments are basically two ratios formed by segment lengths that are equal. We are going to show you two types of lines that result in the creation of proportional segments.

The first type of line used to make a proportional segment is parallel to one side of a triangle and intersects the other two sides. When drawn, this line divides the intersected sides proportionally. For example, in Figure 8.13, $\overline{BC}$ intersects $\overline{AD}$ and $\overline{AE}$ and results in the following proportional segments: $\dfrac{AB}{BD} = \dfrac{AC}{CE}$

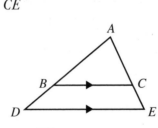

Figure 8.13: $\overline{BC}$ *is a parallel segment in* $\triangle ADE$.

Based on $\triangle ADE$, the following proportion can be written:

$$\frac{AB}{AD} = \frac{AC}{AE}$$

Let's look at an example. In Figure 8.14, with MN as the parallel segment, find LN.

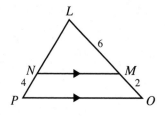

Figure 8.14: $\overline{MN}$ *is a parallel segment in the triangle.*

Because $\overline{MN}$ is parallel to $\overline{OP}$, you can write the following proportion:

$$\frac{LN}{LP} = \frac{LM}{LO}$$

Substitute the numbers from the figure into the proportion, using x to represent LN:

$$\frac{x}{x+4} = \frac{6}{8}$$

$8x = 6(x + 4)$

$8x = 6x + 24$

$8x - 6x = 6x - 6x + 24$

$2x = 24$

$x = 12$

HELPFUL HINT

When solving a proportion, the first step is to cross-multiply.

The second type of line that creates proportional segments is an angle bisector. As stated earlier, an angle bisector is a ray that bisects one angle of a triangle. The ray divides the opposite side into segments that are proportional to the other two sides of the triangle. For example, in Figure 8.13, $\overline{WY}$ is an angle bisector.

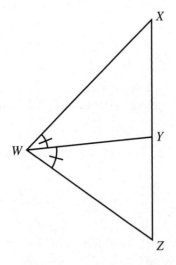

Figure 8.15: *$\overline{WY}$ is an angle bisector of $\triangle WXZ$.*

Based on $\triangle WXZ$, the following proportions can be written: $\dfrac{XY}{WX} = \dfrac{ZY}{WZ}$ and $\dfrac{WX}{WZ} = \dfrac{XY}{ZY}$.

Let's look at an example. Using the previous figure, if $WX = 10$ inches, $XY = 7$ inches, and $WZ = 9$ inches, find ZY.

For $\triangle WXZ$, you can use the proportion $\dfrac{XY}{WX} = \dfrac{ZY}{WZ}$ and substitute in the information, using x to represent ZY:

$$\frac{7}{10} = \frac{x}{9}$$

$$10x = 63$$

$$\frac{10x}{10} = \frac{63}{10}$$

$$x = 6.3 \text{ inches}$$

Inequality Theorems

There are four inequality theorems associated with triangles, each of which states a necessary relationship between the sides and angles of a triangle: the Triangle Inequality Theorem, the Side Angle Inequality Theorem, the Exterior Angle Inequality Theorem, and the Hinge Theorem.

Triangle Inequality Theorem

According to this theorem, the sum of the lengths of any two sides of a triangle must be greater than the third side. For example, a triangle whose sides measure 3, 4, and 6 satisfies the following three inequalities:

$$3 + 4 > 6$$

$$3 + 6 > 4$$

$$4 + 6 > 3$$

Because the sum of the lengths of any two sides of the triangle is greater than the length of the third side, it is proven that these side lengths do form a triangle.

Let's look at an example with side lengths given that are not going to form a triangle. How do you prove that the side lengths 5, 7, and 13 do not form a triangle? In this case, the side lengths 5 and 7 add to a sum of 12. Because 12 is not greater than 13, these side lengths can't form a triangle.

To think of this visually, imagine you have three sticks that measure 5 inches, 7 inches, and 13 inches. If you formed an angle with the two sticks measuring 5 inches and 7 inches, it would be impossible to connect them with the 13-inch stick—the 13-inch stick would be too long.

Let's now try an example that applies the Triangle Inequality Theorem. If two sides of a triangle measure 10 centimeters and 13 centimeters, what are the possible lengths of the third side?

As you know, the sum of the lengths of any two sides of a triangle must be greater than the third side according to this theorem. Using x to represent the length of the unknown side, the following three inequalities would have to be true:

$$10 + x > 13 \qquad 10 + 13 > x \qquad x + 13 > 10$$

inequalities to find the possibilities for the third side length.

owing:

following:

llowing:

$$x + 13 > 10$$

$$x + 13 - 13 > 10 - 13$$

$$x > -3$$

> **IT DOESN'T ADD UP**
>
> Because side lengths can't be negative, the inequality $x > -3$ is not relevant to the answer.

To determine the possibilities for the length of the third side, look at the inequalities $x > 3$ and $23 < x$. According to those, for a triangle with sides of 10 and 13, the length of the third side must be greater than 3 and less than 23 in order to make a proper triangle.

Side Angle Inequality Theorem

Think about an angle as it gets larger and larger; the segment needed to attach the two rays would have to get longer and longer. Therefore, in a triangle, the largest angle is across from the longest side. Likewise, the smallest angle is across from the shortest side. This is known as the Side Angle Inequality Theorem.

For example, if you take a look at Figure 8.16, $\overline{AC}$ is the longest side because it is across from the largest angle, $\angle B$, and $\overline{BC}$ is the shortest side because it is across from the smallest angle, $\angle A$.

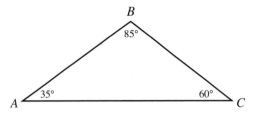

Figure 8.16: *For $\triangle ABC$, you can see the angles and their relation of the size of the side across from them.*

Let's take a look at an example to make sure you understand. In Figure 8.17, determine which side of the triangle is the shortest.

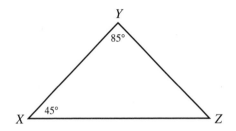

Figure 8.17: *For the triangle, two of the interior angle measures are given.*

The Side Angle Inequality Theorem states that the shortest side is across from the smallest angle. To determine the smallest angle, you must find the measure of the missing angle in the figure. Because the sum of the interior angles of a triangle is 180°, you can solve for the third angle using the following equation:

$$45° + 85° + \angle Z = 180°$$

$$130° + \angle Z = 180°$$

$$130° - 130° + \angle Z = 180° - 130°$$

$$\angle Z = 50°$$

Now that you have the measure of all three angles, you know that $\angle X$ is the smallest angle. The side across from $\angle X$ is $\overline{YZ}$, meaning $\overline{YZ}$ is the shortest side of the triangle.

Exterior Angle Inequality Theorem

As you know, the sum of the interior angles of a triangle always equals 180°. On the exterior of the triangle, the sum of the angles is 360°. Each exterior angle is equal to the sum of the two *remote interior angles.*

DEFINITION

The **remote interior angles** are the two angles not adjacent to the exterior angle in question.

According to the Exterior Angle Inequality Theorem, the measure of an exterior angle of a triangle is greater than the measure of either remote interior angle. This must be true since the sum of the two remote interior angles is equal to the exterior angle. For example, take a look at Figure 8.18.

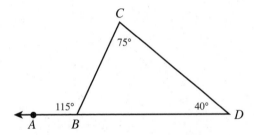

Figure 8.18: *In* $\triangle BCD$, $\angle ABC > \angle BCD$ *and* $\angle ABC > \angle BDC$.

Both interior angles are less than the exterior angle of 115°, proving the theorem.

Let's try using what you've learned with the information in Figure 8.18 to determine the measure of the exterior angle formed at vertex C. The measure of the exterior angle formed at vertex C is equal to the sum of its remote interior angles: $\angle CBD$ and $\angle CDB$. Because the sum of the interior angles of a triangle equals 180°, $m\angle CBD = 65°$.

The sum of the remote interior angles is 105°; therefore, the exterior angle formed at vertex C measures 105°.

Hinge Theorem

According to this theorem, if two sides of one triangle are congruent to two sides of another triangle and the included angle of the first triangle is larger than that of the second, then the third side of the first triangle is longer than the third side of the second triangle. This might sound confusing, but remember when we talked about an angle getting larger and larger? As the angle gets larger, the segment that connects the two segments of the angle must get longer and longer. So if you take two segments and form a 60° angle, the length of the segment that connects them must be longer than the one used to create a 40° angle, as you can see in Figure 8.19.

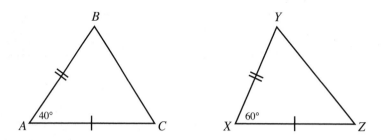

Figure 8.19: *The two triangles have two congruent sides but different angle sizes, meaning YZ and BC are different.*

Because $AC = XZ$, $AB = XY$, and $m\angle X > m\angle A$, $YZ > BC$.

Now that you know the Hinge Theorem, let's use it to compare WX and XY in Figure 8.20.

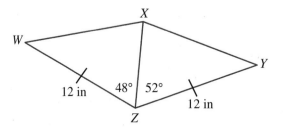

Figure 8.20: $\angle WXZ$ and $\angle YXZ$ attached at $\overline{XZ}$.

In Figure 8.20, $\overline{WZ} \cong \overline{YZ}$ and $\angle YZX > \angle WZX$. By the reflexive property of equality, $\overline{XZ} \cong \overline{XZ}$. Therefore, by the Hinge Theorem, $m\angle WZX < m\angle YZX$, so $WX < XY$.

The Least You Need to Know

* Triangles are classified by their sides (equilateral, isosceles, and scalene) and angles (acute, obtuse, right, and equiangular).

* Altitudes, medians, angle bisectors, and perpendicular bisectors are concurrent lines or segments that meet at a point of concurrency. These can help you find the missing measures of the sides or angles of triangles.

* The midsegment connects the midpoint of two sides of a triangle. It is parallel to and half the length of the third side of the triangle.

* You can use proportional segments to solve for the length of a missing side of a triangle.

* Inequality theorems state the relationships between sides and angles that must be true in order to create a triangle.

Right Triangles and Trigonometry

As you saw in the previous chapter, while all triangles have three sides and an interior angle sum of 180°, they are not the same. Right triangles, in particular, have many special math concepts, such as the Pythagorean Theorem, special right triangles, and trigonometric ratios. In this chapter, we look more deeply into what makes right triangles stand out from the rest of the triangles.

The Pythagorean Theorem

The Pythagorean Theorem, named after the Greek mathematician Pythagoras (ca. 570–495 B.C.E.), states that in a right triangle, the square of the *hypotenuse* is equal to the sum of the squares of the legs. Let's explore this theorem using Figure 9.1.

In This Chapter

- How the Pythagorean Theorem can help you find the missing side of a right triangle
- Finding the type of triangle with the converse of the Pythagorean Theorem
- Special right triangles: 30°-60°-90° and 45°-45°-90°
- Using trigonometric ratios to find the heights of objects too tall to measure

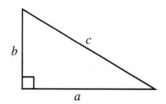

DEFINITION

The **hypotenuse** is the longest side of a right triangle. It is always across from the right angle.

Figure 9.1: *A right triangle with sides labeled a, b, and c.*

Obviously, the triangle is a right triangle because it has a right angle. The side labeled c is the hypotenuse because it is across from the right angle, while the sides labeled a and b are the legs. Based on this right triangle, you can write an equation using the Pythagorean Theorem as follows:

$$a^2 + b^2 = c^2$$

Even if you didn't recognize the term, you probably know this equation pretty well. Let's see the equation in action. For Figure 9.2, find the length of the missing side.

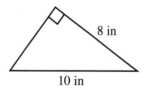

8 in

10 in

Figure 9.2: *In this right triangle, the length of one leg and the hypotenuse are given.*

Because the triangle is a right triangle, the Pythagorean Theorem can be used to determine the length of the missing side. The side with length 10 inches is across from the right angle, so it is the hypotenuse. The side that measures 8 inches is one of the legs. You can plug these values into the equation and solve for the missing side:

$$a^2 + b^2 = c^2$$

$$a^2 + 8^2 = 10^2$$

$$a^2 + 64 = 100$$

$$a^2 + 64 - 64 = 100 - 64$$

$a^2 = 36$

$\sqrt{a^2} = \sqrt{36}$

$a = 6$ inches

> **HELPFUL POINT**
>
> When plugging in the value for one leg, it doesn't matter if you put it in for a or b; the answer will be the same.

Let's try another example. Did you know that a television is measured by the length of its diagonal? The diagonal of a television separates the rectangular television into two right triangles—the length and width of the television become the legs, and the diagonal is the hypotenuse, as you can see in Figure 9.3.

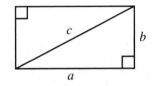

Figure 9.3: *The diagonal in the rectangle is the hypotenuse of the right triangles.*

Say the figure represents a 50-inch television, meaning the diagonal (labeled c) measures 50 inches. If the length of the television is 40 inches, find the width of the television.

Because the diagonal forms a right triangle, you can plug in the values to the Pythagorean Theorem equation. The diagonal, 50 inches, is the hypotenuse and the length, 40 inches, is one of the legs. Solve for the missing side:

$a^2 + b^2 = c^2$

$40^2 + b^2 = 50^2$

$1600 + b^2 = 2500$

$1600 - 1600 + b^2 = 2500 - 1600$

$b^2 = 900$

$\sqrt{b^2} = \sqrt{900}$

$b = 30$ inches

So the width of the television is 30 inches.

Pythagorean Triples

A *Pythagorean triple* is a right triangle whose sides are in the ratio 3:4:5. The side lengths do not need to measure 3, 4, and 5; however, they do need to reduce to that ratio. For example, the side lengths 9, 12, and 15 are a Pythagorean triple because the ratio 9:12:15 simplifies to the ratio 3:4:5. Any set of three side lengths that simplifies to the ratio 3:4:5 is a right triangle.

Here's another example: Would the side lengths 27, 36, and 44 be considered a Pythagorean triple? To find out, you have to figure out whether there's a common factor for all three numbers that reduces them to 3, 4, and 5, respectively. In this case, while 27 and 36 have the common factor of 9 and can be simplified to the ratio 3:4, the length of 44 is not divisible by 9. Therefore, the side lengths can't be simplified into the ratio 3:4:5, meaning the triangle is not a Pythagorean triple.

The Converse of the Pythagorean Theorem

You can use the Pythagorean Theorem for more than finding side lengths in right triangles. You can also use it to determine if a triangle is acute, right, or obtuse. This is known as the converse of the Pythagorean Theorem, which reads as follows:

If $c^2 < a^2 + b^2$, then the triangle is acute.

If $c^2 = a^2 + b^2$, then the triangle is right.

If $c^2 > a^2 + b^2$, then the triangle is obtuse.

IT DOESN'T ADD UP

Don't confuse your inequality symbols! The less-than symbol is < and the greater-than symbol is >.

For example, if a triangle has side lengths 6 cm, 7 cm, and 8 cm, you can plug it into the equation to determine if the triangle is acute, right, or obtuse. Because the hypotenuse is the longest side, you have to make the assumption that $c = 8$. You can then use $a = 6$ cm and $b = 7$ cm, though the legs are interchangeable. Now solve to see if the sum of the squares of the legs is less than, equal to, or greater than the square of the hypotenuse:

c^2 ___ $a^2 + b^2$

8^2 ___ $6^2 + 7^2$

64 ___ $36 + 49$

$64 < 85$

Because the square of the hypotenuse is less than the sum of the squares of the legs, the triangle is acute.

Altitude Drawn to the Hypotenuse

If you recall from Chapter 8, an altitude is the height of the triangle. To draw an altitude to the hypotenuse, you begin at the right angle and draw a line that is perpendicular to the hypotenuse. In right triangle $\triangle WXY$, the altitude $\overline{WZ}$ is drawn to the hypotenuse $\overline{XY}$.

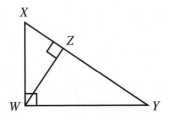

Figure 9.4: *The altitude separates right triangle $\triangle WXY$ into two triangles.*

In Figure 9.4, the original triangle is split into two new triangles. You have three triangles in total—the triangle as a whole and the two small triangles. The smaller triangles formed by the altitude are called *similar triangles* because they are similar to each other and similar to the original triangle.

We will explore similar triangles more in depth in Chapter 13. For now, let's look at how the similarity formed by an altitude can help to determine side lengths. Because the triangles are similar, the following theorems can be stated in regard to the triangle in Figure 9.4:

If an altitude is drawn to the hypotenuse of a right triangle, then each leg is the *geometric mean* between the hypotenuse and its adjacent segment on the hypotenuse:

$$\frac{XY}{XW} = \frac{XW}{XZ}$$

and

$$\frac{YX}{YW} = \frac{YW}{YZ}$$

If an altitude is drawn to the hypotenuse of a right triangle, then the altitude is the geometric mean between the segments on the hypotenuse:

$$\frac{XZ}{WZ} = \frac{WZ}{YZ}$$

DEFINITION

Similar triangles are different sizes of the same-shaped triangle. The angles are the same and the side lengths are proportional.

Geometric mean is the positive number x, such that $\dfrac{a}{x} = \dfrac{x}{b}$.

Let's take a look at how this works. Using Figure 9.5, find the value of x and y.

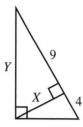

Figure 9.5: *The altitude separates the triangle into two similar triangles, with missing values x and y.*

To solve, you first need to determine which proportion you need to use for x based on the proportions devised for Figure 9.4. Because x is the altitude, you need the proportion that uses the altitude. From looking at ΔWXY, you can see you need to use $\dfrac{XZ}{WZ} = \dfrac{WZ}{YZ}$ and relate it to the new triangle in Figure 9.5. Here's how you substitute the values into the proportion:

$$\frac{9}{x} = \frac{x}{4}$$

$$9(4) = x(x)$$

$$36 = x^2$$

$$\sqrt{36} = \sqrt{x^2}$$

$$6 = x$$

Now you need to determine the proportion to use to find y. Because y is one of the legs, you can use $\dfrac{XY}{XW} = \dfrac{XW}{XZ}$ and relate it to the triangle in Figure 9.5. Here's how you substitute the values into the proportion for this one:

$$\frac{13}{y} = \frac{y}{9}$$

$$13(9) = y(y)$$

$$117 = y^2$$

$$\sqrt{117} = \sqrt{y^2}$$

$$\sqrt{117} = y$$

$\sqrt{117}$ results in an irrational number and does not have a factor that is a perfect square, so you just leave it in this form.

Special Right Triangles

Two types of right triangles are considered special right triangles because the ratio of their side lengths is always the same. The size of the triangle does not matter; it just needs to have specific measures for its angles. One of the special right triangles has angles that measure 30°, 60°, and 90°. The other special right triangle has angles that measure 45°, 45°, and 90°.

30°-60°-90° Triangle

The lengths of the sides of a 30°-60°-90° triangle are in a ratio of $1 : \sqrt{3} : 2$, which you can see in Figure 9.6.

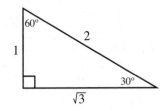

Figure 9.6: *This triangle is a 30°-60°-90° triangle with the ratio of its side lengths labeled.*

In the figure, the 2 describes the hypotenuse, the 1 describes the leg across from the 30° angle, and $\sqrt{3}$ describes the leg across from the 60° angle.

HELPFUL POINT

If you recall from Chapter 8, the shortest side is always across from the smallest angle and the longest side is always across from the largest angle.

Now that you know the setup of a 30°-60°-90° triangle, let's try an example. Determine the value of x and y for the triangle in Figure 9.7.

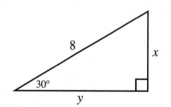

Figure 9.7: *A 30°-60°-90° triangle with missing sides x and y.*

In the triangle, the hypotenuse is 8. Based on the ratio of the side lengths in a 30°-60°-90° triangle, the hypotenuse is twice the length of the side opposite the 30° angle. Therefore, x must be the following:

$$\left(\frac{1}{2}\right)(8) = 4$$

Next, you find the value of y. The length of the side opposite the 60° angle is found by multiplying the length of the shortest side by $\sqrt{3}$. Thus, the length of the side opposite the 60° angle is $4\sqrt{3}$ because it can't be reduced further.

To help solve more difficult problems, it is useful to label values for each of the three sides in a 30°-60°-90° triangle. You label the shorter leg a, the longer leg $a\sqrt{3}$, and the hypotenuse $2a$. Using this information, determine the value of x and y for the triangle in Figure 9.8.

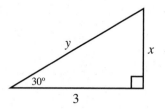

Figure 9.8: *A 30°-60°-90° triangle with missing sides x and y.*

Using the preceding figure, $a = x$, $a\sqrt{3} = 3$, and $2a = y$, so three equations have been created from the triangle.

The second triangle has only one variable, so you can solve this equation for a:

$$a\sqrt{3} = 3$$

$$\frac{a\sqrt{3}}{\sqrt{3}} = \frac{3}{\sqrt{3}}$$

$$a = \frac{3}{\sqrt{3}} \bullet \frac{\sqrt{3}}{\sqrt{3}}$$

$$a = \frac{3\sqrt{3}}{3}$$

$$a = \sqrt{3}$$

Now you can use the value of a to substitute into the two other equations. First, substitute into $a = x$ to find the value of x:

$$a = x$$

$$\sqrt{3} = x$$

Finally, substitute into $2a = y$ to find the value of y:

$$2a = y$$

$$2\sqrt{3} = y$$

45°-45°-90° Triangle

The lengths of the sides of a 45°-45°-90° triangle are in a ratio of $1 : 1 : \sqrt{2}$, as shown in Figure 9.9.

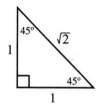

Figure 9.9: *This triangle is a 45°-45°-90° triangle with the ratio of its side lengths labeled.*

HELPFUL POINT

A 45°-45°-90° triangle has two sides of equal length; therefore, it is an isosceles triangle.

To help solve problems with 45°-45°-90° triangles, you can label each of the sides based on the ratio. Each of the legs will be labeled a and the hypotenuse will be labeled $a\sqrt{2}$.

Let's now take a look at an example. Given a right triangle with the two legs each measuring 5 inches as in Figure 9.10, determine the length of the hypotenuse.

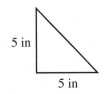

5 in

5 in

Figure 9.10: *A right triangle with side lengths of 5 inches.*

Because this is a right triangle with two equal sides, you know it is a 45°-45°-90° triangle. The legs measure 5 inches; therefore, $a = 5$. Because $a = 5$, $a\sqrt{2} = 5\sqrt{2}$. So the length of the hypotenuse is $5\sqrt{2}$.

Trigonometric Ratios

Trigonometric ratios are often used to solve for the length of objects that are too long to measure and involve right triangles. This could be the height of a skyscraper, the height of a tree, or the height of a telephone pole.

Lots of times, math teachers love to come up with acronyms for math concepts. (Remember PEMDAS for the order of operations?) Well, trigonometric ratios are another one of those concepts with an acronym: SOHCAHTOA. SOHCAHTOA is an acronym for Sine is Opposite over Hypotenuse, Cosine is Adjacent over Hypotenuse, and Tangent is Opposite over Adjacent. It probably sounds silly, but it works!

Sine, cosine, and tangent are three of the trigonometric ratios. Their ratios are formed by the side lengths of the triangle: adjacent leg is the leg adjacent to the angle used, opposite leg is the leg opposite the angle used, and hypotenuse is the side that is the hypotenuse of the right triangle.

For example, consider the $\triangle ABC$ in Figure 9.11.

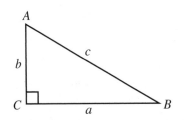

Figure 9.11: *A right triangle with sides and angles labeled.*

The sine of angle A, abbreviated $\sin \angle A$, equals the ratio of the opposite leg to the hypotenuse. Using the preceding triangle, this can be expressed as $\sin \angle A = \dfrac{a}{c}$.

The cosine of angle A, abbreviated $\cos \angle A$, equals the ratio of the adjacent leg to the hypotenuse. Using the preceding triangle, this can be expressed as $\cos \angle A = \dfrac{b}{c}$.

The tangent of angle A, abbreviated $\tan \angle A$, equals the ratio of the opposite leg to the adjacent leg. Using the preceding triangle, this can be expressed as $\tan \angle A = \dfrac{a}{b}$.

Now that you have these ratios, let's try an example. A telephone pole is supported by a wire extending from the top of the pole to a metal stake in the ground. The wire is 26 feet long and forms a 60° angle with the ground. How tall is the telephone pole?

Let x represent the height of the telephone pole, as shown in Figure 9.12, and solve:

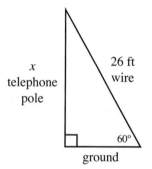

Figure 9.12: *The figure represents a telephone pole and the wire supporting it.*

$$\cos 60° = \frac{b}{26}$$

$$0.5 = \frac{b}{26}$$

$$0.5(26) = \frac{b}{26}(26)$$

$$13 = b$$

So the telephone pole is approximately 13 feet tall.

HELPFUL POINT

Whenever possible, draw a picture. Drawing a picture is a great strategy to solve problems like those for determining the length of very long objects.

The Least You Need to Know

- The Pythagorean Theorem can be used to find missing side lengths in right triangles.
- The converse of the Pythagorean Theorem helps you classify a triangle by its angles.
- When an altitude is drawn to the hypotenuse of a right triangle, it forms similar triangles.
- There are two types of special right triangles: 30°-60°-90° and 45°-45°-90°. Their side lengths are in a constant ratio.
- Trigonometric ratios can be used to find missing angle measures and missing side lengths in right triangles. This is especially helpful when trying to find the height of something too tall to measure manually.

CHAPTER 10

Congruent Triangles

How do you know if two things are exactly the same? To be exactly the same, all the parts must be the same. When it comes to triangles, this means the sides and angles of two triangles must be the same. In this chapter, we tell you how to find the corresponding parts of triangles, use postulates and theorems to prove two triangles are congruent, and fill out proofs that prove the triangles and the parts of triangles are congruent.

In This Chapter

* Identifying corresponding sides and angles of triangles
* Determining if triangles are congruent
* Formal proofs of congruent triangles

Corresponding Sides and Angles of Two Triangles

In two congruent triangles, all of the parts of one triangle are congruent to the *corresponding parts* of the other triangle. The corresponding parts include the three sides and the three angles, as you can see in Figure 10.1.

DEFINITION

For two triangles, the **corresponding parts** of two triangles are the matching sides or angles.

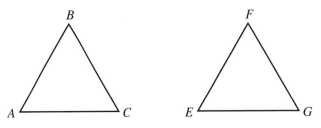

Figure 10.1: *These two triangles are congruent.*

In the preceding triangles, $\overline{AB}$ corresponds to $\overline{EF}$, $\overline{AC}$ corresponds to $\overline{EG}$, and $\overline{BC}$ corresponds to $\overline{FG}$.

There are also three corresponding angles: $\angle A$ corresponds to $\angle E$, $\angle B$ corresponds to $\angle F$, and $\angle C$ corresponds to $\angle G$.

When it comes to naming congruent triangles, it is important that the corresponding parts are in the same order. For example, the triangles in Figure 10.1 would have to be named $\triangle ABC$ and $\triangle EFG$.

Let's take a look at another example in which we have only the name of two congruent triangles, $\triangle LMN$ and $\triangle QRS$. Even with just the names for these triangles, you can still identify the corresponding angles and sides without seeing images of them:

$\angle L$ corresponds to $\angle Q$

$\angle M$ corresponds to $\angle R$

$\angle N$ corresponds to $\angle S$

$\overline{LM}$ corresponds to $\overline{QR}$

$\overline{MN}$ corresponds to $\overline{RS}$

$\overline{LN}$ corresponds to $\overline{QS}$

Postulates and Theorems for Congruent Triangles

As you know, two triangles are congruent if their corresponding sides and angles are congruent. It could be a lot of work to show that all six measurements are congruent. However, there are four postulates and one theorem that lessen the work. For a pair of triangles, certain groups of three of these measurements can prove their congruency.

The following are the postulates and theorem that can help you identify whether two triangles are congruent:

All three corresponding sides are equal (SSS Postulate). Short for "side-side-side," this postulate says if the three sides can be proven congruent, the angles that attach these sides must also be congruent. With a given set of three side lengths, there is only one unique triangle that can be drawn.

> **HELPFUL POINT**
>
> A triangle is unique when there is not another triangle that has its exact dimensions or shape. To help you understand, two triangles with the same angle measures do not have to be equivalent because the side lengths can vary. Therefore, a triangle with angles measuring 30°, 40°, and 110° is not necessarily unique.

Two corresponding sides and their included angle are equal (SAS Postulate). Short for "side-angle-side," this postulate says, if two sides are attached at the same angle measure, the side that attaches their two endpoints must be congruent.

Two corresponding angles and their included side are equal (ASA Postulate). Short for "angle-side-angle," this postulate says, in a pair of congruent triangles, if two corresponding angles are congruent and connected by a congruent side, the two other sides must be congruent.

Two corresponding angles and a nonincluded side are equal (AAS Postulate). Short for "angle-angle-side," this postulate says, when two angles in a triangle are congruent, the third angle must also be congruent because the sum of the interior angles of a triangle is 180°. When one segment is proven congruent along with those angles, the other two sides must be congruent.

The hypotenuse and one corresponding leg of two right triangles are equal (HL Theorem). Short for "hypotenuse-leg," this theorem says, in a right triangle, if the hypotenuse and a leg are congruent, the other leg must also be congruent. Just think about it in terms of the Pythagorean theorem—to make the equation true, there is only one value left to plug in for the missing leg length.

⊕✕ **IT DOESN'T ADD UP**

There is no SSA Postulate because an angle and two sides do not guarantee two triangles are congruent. Don't make this mistake!

Let's take a look at some triangles and see which postulate or theorem applies to them.

In Figure 10.2, you have two right triangles. Which postulate or theorem applies?

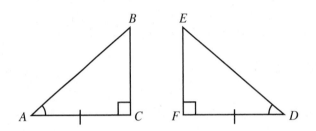

Figure 10.2: *Both triangles are right triangles with additional congruent parts.*

Because △*ABC* and △*DEF* are both right triangles, they already have one pair of corresponding congruent angles. And by the markings on the figures, $\overline{AC} \cong \overline{DF}$ and $\angle A \cong \angle D$. Therefore, the two triangles are proven congruent by the ASA Postulate because two angles and their included side are congruent.

Figure 10.3 shows another set of triangles. Try to figure out what postulate or theorem applies to them.

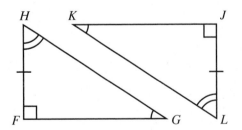

Figure 10.3: *These triangles have three pairs of corresponding congruent parts.*

△*FGH* and △*JKL* have one pair of corresponding congruent sides. Based on the congruency markings, $\overline{FH} \cong \overline{JL}$. There are also two pairs of corresponding congruent angles, $\angle G \cong \angle K$ and $\angle H \cong \angle L$, because the congruency markings match between the two triangles. Therefore, the two triangles are proven congruent by the AAS Postulate because two pairs of corresponding angles and a pair of corresponding nonincluded sides are congruent.

Try one more. Which postulate or theorem applies to Figure 10.4?

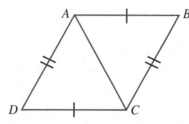

Figure 10.4: *These triangles share a common side.*

△*ACD* and △*CAB* are attached by a common side. Because the side is shared, it must be congruent to itself by the reflexive axiom of equality (see Chapter 6). By the congruency markings on the figure, $\overline{CD} \cong \overline{AB}$ and $\overline{DA} \cong \overline{BC}$. Therefore, the triangles are congruent by the SSS Postulate because there are three pairs of corresponding congruent sides.

Proving Triangles Are Congruent

In Chapter 6, you learned the basics of writing a two-column proof. Here, we are going to expand on those by writing proofs to prove that two triangles are congruent.

Take a look at the following proof with its given information.

Given: $\overline{AC} \cong \overline{BC}$ and $\overline{CM} \perp \overline{AB}$

Prove: △*ACM* ≅ △*BCM*

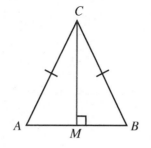

Figure 10.5: *An isosceles triangle with height drawn.*

Statements	Reasons
1. $\overline{AC} \cong \overline{BC}$ and $\overline{CM} \perp \overline{AB}$	1. Given
2. $\angle CMA$ and $\angle CMB$ are right angles	2. Definition of perpendicular
3. $m\angle CMA = 90°$ and $m\angle CMB = 90°$	3. Definition of right angle
4. $m\angle AMC = m\angle BMC$	4. Substitution property of equality
5. $\angle AMC \cong \angle BMC$	5. Definition of congruence
6. $\overline{CM} \cong \overline{CM}$	6. Reflexive axiom of equality
7. $\triangle ACM \cong \triangle BCM$	7. HL Theorem

Using the figure and the given information, you can see that $\overline{CM}$ intersects $\overline{AB}$ and forms a right angle. This means these two segments are perpendicular and result in $\angle AMC$ and $\angle BMC$ being right angles, which measure 90°. The substitution property allows you to state that $\angle AMC$ and $\angle BMC$ are congruent. The figure also shows that the two triangles share $\overline{CM}$. Because the two triangles are right triangles, one pair of legs are congruent, and their hypotenuses are congruent, the triangles are congruent by the HL Theorem.

HELPFUL POINT

When completing proofs of congruent triangles, it is useful to mark congruent sides and angles on the figure. It helps in identifying the correct triangle congruency postulate or theorem.

Proofs can be intense, but with practice, they become easier. Because practice is a good way to get better at proofs, let's try another example.

Given: $\overline{AB}$ and $\overline{CD}$ bisect at M

Prove: $\triangle AMD \cong \triangle BMC$

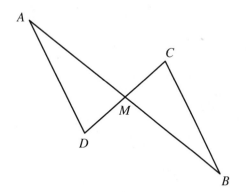

Figure 10.6: *Two triangles that share vertex M.*

Statements	Reasons
1. $\overline{AB}$ and $\overline{CD}$	1. Given
2. $\overline{AM} \cong \overline{BM}$ and $\overline{MD} \cong \overline{MC}$	2. Definition of segment bisector
3. $\angle AMD \cong \angle BMC$	3. Vertical Angles Congruency Theorem
4. $\triangle AMD \cong \triangle BMC$	4. SAS Postulate

The definition of segment bisector states that when two segments bisect, they form congruent segments; therefore, $\overline{AM} \cong \overline{BM}$ and $\overline{MD} \cong \overline{MC}$. Looking at the intersection of $\overline{AB}$ and $\overline{CD}$, the angles formed are vertical angles; the Vertical Angles Congruency Theorem states that vertical angles are congruent. Because there are two pairs of corresponding congruent sides and their included angles are congruent, the triangles are congruent by the SAS Postulate.

Proving Corresponding Parts of Congruent Triangles Are Congruent

When triangles are proven congruent, it is also proven that their corresponding parts are congruent. The following proofs work to prove that corresponding parts of congruent triangles are congruent (abbreviated as CPCTC).

👉 **HELPFUL POINT**

Before proving corresponding parts of triangles congruent, you must prove the triangles congruent, as you learned in the previous section.

When completing proofs with CPCTC, the process follows very similarly to the previous proofs in this chapter. Let's take a look!

Given: $\overline{MR} \perp \overline{RP}$ and $\overline{QP} \perp \overline{RP}$; O is the midpoint of $\overline{RP}$

Prove: $\angle M \cong \angle Q$

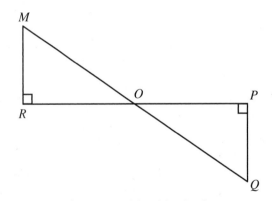

Figure 10.7: *Two right triangles.*

Statements	Reasons
1. $\overline{MR} \perp \overline{RP}$ and $\overline{QP} \perp \overline{RP}$	1. Given
2. $\angle MRO$ and $\angle QPO$ are right angles	2. Definition of perpendicular
3. $m\angle MRO = 90°$ $m\angle QPO = 90°$	3. Definition of right angles
4. $m\angle MRO = m\angle QPO$	4. Substitution property of equality
5. $\angle MRO \cong \angle QPO$	5. Definition of congruence
6. O is the midpoint of $\overline{RP}$	6. Given
7. $\overline{OR} \cong \overline{OP}$	7. Definition of midpoint
8. $\angle MOR \cong \angle QOP$	8. Vertical Angles Congruency Theorem
9. $\triangle MOR \cong \triangle QOP$	9. ASA Postulate
10. $\angle M \cong \angle Q$	10. CPCTC

Whenever segments are perpendicular, the intersection of these segments form right angles by the definition of *perpendicular*. These right angles measure 90° and are congruent. Using additional information from the given statement, $\overline{RO} \cong \overline{PO}$ because the midpoint of a segment separates the segment into two equal segments. Aside from the given information, the intersection of MQ and RP forms vertical angles. By the Vertical Angles Congruency Theorem, $\angle MOR \cong \angle QPO$. Because the triangles have two pairs of corresponding congruent angles and their included sides are congruent, the triangles are congruent by the ASA Postulate. CPCTC can then be used to prove any of the corresponding parts are congruent.

In some of the preceding examples, the triangles have been attached on a side or at a vertex. However, there are also figures where the triangles are not just attached on a side or at a vertex; they are partially on top of each other. These triangles are called *overlapping triangles*. Figure 10.8 shows a pair of overlapping triangles.

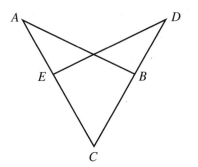

Figure 10.8: *Two triangles that are overlapping.*

When you are dealing with overlapping triangles, it is useful to separate the two triangles and redraw them. Figure 10.9 shows the separated triangles.

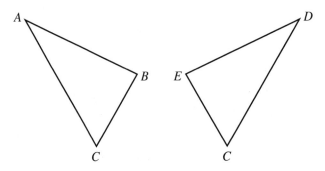

Figure 10.9: *The two overlapping triangles have been separated.*

As you can see, the overlapping triangles are $\triangle ABC$ and $\triangle DEC$.

 **HELPFUL POINT**

Notice that the two triangles share vertex C. All overlapping triangles will have common points.

Now that the triangles have been separated, let's look at a proof using Figure 10.9.

Given: $\angle ABC \cong \angle DEC$ and $\overline{CB} \cong \overline{CE}$

Prove: $\overline{AB} \cong \overline{DE}$

Statements	Reasons
1. $\angle ABC \cong \angle DEC$ and $\overline{CB} \cong \overline{CE}$	1. Given
2. $\angle ACB \cong \angle DCE$	2. Reflexive axiom of equality
3. $\triangle ABC \cong \triangle DEC$	3. ASA Postulate
4. $\overline{AB} \cong \overline{DE}$	4. CPCTC

When the two triangles become separated, it can be seen that both triangles contain $\angle C$. The reflexive axiom of equality states that these angles are congruent. With the given information and $\angle C \cong \angle C$, the two triangles are congruent by the ASA Postulate. CPCTC can then be used to prove any of the corresponding parts are congruent.

The Least You Need to Know

- Triangles are congruent if all their corresponding parts (the three sides and three angles) are congruent.
- There are postulates and a theorem that can be used to prove triangles congruent using only three pairs of corresponding congruent parts: SSS, SAS, ASA, AAS, and HL.
- Formal proofs can be written to show that triangles, as well as their corresponding parts, are congruent.

Two-Dimensional Figures

This part focuses on the elements of the world around you that are two-dimensional. You take an in-depth look at the relationships that exist between the side lengths and angles of several two-dimensional figures classified as polygons. The characteristics of four-sided polygons are then more specifically discussed and used to write proofs. Theorems related to similar triangles are also explored to strengthen your understanding of ratios and proportions.

Polygons

Now that you've learned about triangles, let's extend those properties to look at other polygons that are closed-plane figures with straight lines and greater than three sides. In Euclidean geometry, these figures are primarily identified by the number of sides, so in this chapter, we help you explore relationships among a polygon's sides and angles.

The most important measures of polygons are perimeter and area, which can be applied to real-life situations. We take a look at some of these situations here, as well as ways to find these measures for figures with greater than three sides.

In This Chapter

- Identifying and describing polygons
- Finding the measures of the interior and exterior angles of polygons
- Understanding and determining the perimeter and area of polygons

Classifying Polygons

Different shapes have different names. In the case of *polygons,* which all have straight sides, you can more specifically identify them based on the number of sides, side lengths, angles measures, and other characteristics. Let's take a closer look.

DEFINITION

A **polygon** is two-dimensional, closed shape with straight lines. Each line segment in a polygon connects to two other line segments at its endpoints.

Characteristics of Polygons

To better understand polygons, let's try an example. Take a bunch of straws; each straw can be considered a line segment. If you join two straws together, they connect at exactly one point. Can you rearrange the two straws to create a closed shape? The answer is no. You need a third straw. With three straws, you can create a triangle. A triangle is the most basic example of a three-sided polygon. Based on this, can you identify which shapes in Figure 11.1 are polygons?

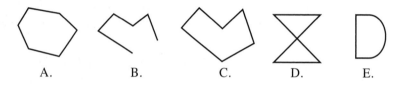

A.   B.   C.   D.   E.

Figure 11.1: *Which of these figures represents a polygon?*

One you can eliminate right away is Figure 11.1B. This figure is not a polygon because it is not closed. One of the main characteristics of a polygon is it must be closed.

Now take a look at Figure 11.1D. This is not a polygon because the vertex of one of the line segments intersects the vertices of three other line segments. Because it connects to more than two, you do not have a polygon.

Figure 11.1E is also not a polygon because it is not made up of straight line segments.

Now that you know which figures are not polygons and why, it is easier to understand that Figures 11.1A and 11.1C are the only polygons. They are closed figures made up of straight line segments, and each line segment is connected to two other line segments at its vertices.

Polygons with *n* Sides

All polygons don't have the same number of sides. In order to differentiate them, we more specifically name polygons based on their number of sides. As mentioned earlier, the simplest polygon is a triangle. Here are the names of polygons with greater than three sides:

- A polygon with four sides is a quadrilateral.

- A polygon with five sides is a pentagon.

- A polygon with six sides is a hexagon.

- A polygon with seven sides is a heptagon.

- A polygon with eight sides is an octagon.

- A polygon with nine sides is a nonagon.

- A polygon with 10 sides is a decagon.

When referring to a polygon with greater than 10 sides, let *n* represent a whole number greater than 10, so a polygon with *n* sides is an *n*-gon. For example, a polygon with 11 sides is known as an 11-gon.

Regular Polygons

What constitutes a "regular" polygon? A polygon is regular if all of its sides are equal (or *equilateral*) and all of its angles are equal (or *equiangular*). For example, a square has equal side lengths and equal angle measures, meaning it is a regular polygon. Another example of a regular polygon is an equilateral triangle. However, if the polygon is not equilateral or equiangular, it is an irregular polygon.

DEFINITION

Equilateral means the sides are equal in length. **Equiangular** means the angle measures are equal.

Figure 11.2 shows a few different polygons. Let's go through their classifications based on both the number of sides and whether it's regular or irregular.

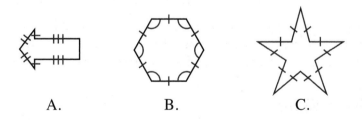

Figure 11.2: *How are these polygons classified?*

Figure 11.2A has seven sides with different side lengths, making it an irregular heptagon. Figure 11.2B is a six-sided equilateral, equiangular polygon; therefore, it is a regular hexagon. Figure 11.2C has 10 sides and is equilateral yet not equiangular. It is classified as an irregular decagon.

Convex vs. Concave

Refer back to Figures 11.1A and 11.1C. Both of them are hexagons, but they don't look alike. What's different about these figures?

Notice how Figure 11.1C "caves" in, while Figure 11.1A does not. Figure 11.1C is an example of a *concave polygon;* Figure 11.1A is an example of a *convex polygon.* A polygon is convex if and only if the measure of each of the interior angles is less than 180°; otherwise, the polygon is concave. You can see another example of a concave polygon in Figure 11.3.

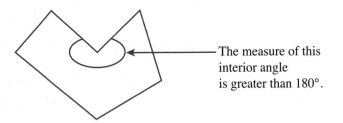

The measure of this
interior angle
is greater than 180°.

Figure 11.3: *Note how this concave polygon "caves" in the center and has angles greater than 180°.*

The following are some important concepts to keep in mind about convex and concave polygons:

- A regular polygon will always be convex. Because a regular polygon is always equiangular, each of its angles must measure less than 180°.

- A triangle will always be a convex polygon. It is not possible for a closed shape to cave in with three sides; the fewest number of sides that a concave polygon can have is four, as you can see in Figure 11.4.

- An irregular polygon can be convex or concave, as you can see in Figure 11.5.

 HELPFUL POINT

If you can connect two points inside the polygon, such that part of the line segment connecting these points is outside of the polygon, the polygon is concave. See Figure 11.5.

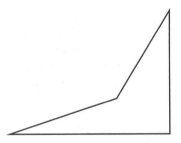

Figure 11.4: *The fewest number of sides that a concave polygon can have is four.*

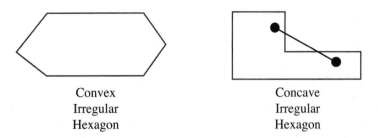

Convex
Irregular
Hexagon

Concave
Irregular
Hexagon

Figure 11.5: *Examples of convex and concave irregular polygons.*

Angle Measures of Polygons

In Chapter 8, you learned that the sum of the measures of the interior angles of a triangle is 180° (known as the *Triangle Sum Theorem*). Let's now investigate the sum of the interior angle measures of polygons with greater than three sides.

The Sum of the Interior Angles of a Convex Polygon

In order to find the sum of the interior angles of a polygon, you need to find out how many diagonals you can draw in it. A diagonal is a line segment that connects two nonconsecutive vertices; it can be drawn from any one of its vertices. Figure 11.6 shows Quadrilateral *ABCD* with a diagonal.

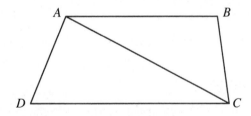

Figure 11.6: *The diagonal of a quadrilateral from one vertex point.*

Quadrilateral *ABCD* has one diagonal from vertex *A*—namely, $\overline{AC}$.

But what does this tell you about the sum of its angles? Notice how drawing all possible diagonals from one of the quadrilateral's vertices forms two triangles. You can use this information to find the sum of the interior angles of a quadrilateral.

The number of triangles formed due to the diagonals times 180° equals the sum of the interior angles. Therefore, the sum of the interior angles of a quadrilateral is the following:

$$2 \times 180° = 360°$$

This is known as the *Quadrilateral Sum Theorem*, which states that the sum of the interior angles of any quadrilateral will always be 360°.

Let's try this with a pentagon. As you can see in Figure 11.7, you can draw two diagonals from vertex A to nonconsecutive vertices C and D—namely, $\overline{AC}$ and $\overline{AD}$. This forms three triangles.

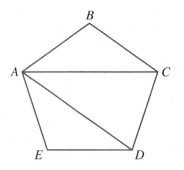

Figure 11.7: *The diagonals of a pentagon.*

Because three triangles are formed, the sum of the interior angles of a pentagon is 3 multiplied by 180°:

$$3 \times 180° = 540°$$

The following chart shows you how to get the sum of the interior angles based on the polygon. As you look at this, compare the number of sides to the number of triangles formed.

Polygon	Number of Sides	Number of Diagonals	Number of Triangles Formed	Sum of the Measures of the Interior Angles
Triangle	3	0	1	1 × 180
Quadrilateral	4	1	2	2 × 180
Pentagon	5	2	3	3 × 180
Hexagon	6	3	4	4 × 180
Heptagon	7	4	5	5 × 180
Octagon	8	5	6	6 × 180
Nonagon	9	6	7	7 × 180
Decagon	10	7	8	8 × 180

Notice how the number of triangles formed is always 2 less than the number of sides!

> **HELPFUL POINT**
>
> The sum of the interior angles of a convex n-gon with n vertices and n sides will always equal $(n - 2) \times 180°$.

If the polygon is a regular polygon, you can find the measure of each of its angles by taking the sum of the interior angles and dividing it by the number of sides. So the measure of each angle in a convex regular polygon is $\dfrac{(n-2) \times 180°}{n}$.

Now that you understand interior angles of a polygon, let's work through some numerical examples.

Suppose you are told to find the sum of the interior angles of a decagon. Because a decagon is 10-sided, the sum of the interior angles of a decagon is the product of 10 subtracted from 2, which is then multiplied by 180°:

$$(10 - 2) \times 180° = 1440°$$

So the sum of the interior angles of a decagon is 1440°.

As another example, suppose you are asked to find the measure of each angle in a *regular* octagon. First, you determine the sum of its interior angles by letting $n = 8$, which is the number of sides in an octagon:

$$(8 - 2) \times 180° = 1080°$$

Because the octagon is regular, you know it is equiangular. Therefore, you divide the sum of the interior angles by the number of angles:

$$1080° \div 8 = 135°$$

So the measure of each angle in a regular octagon is 135°.

To end this discussion on interior angles, let's look at an algebraic example. Figure 11.8 shows a hexagon with one of the angle measures labeled as an algebraic expression. How do you solve for x?

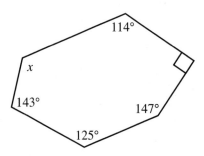

Figure 11.8: *An algebraic application of the sum of the interior angles of a polygon.*

The value of x can be determined by writing an equation where the sum of expressions representing the interior angle measures is equal to the sum of the interior angles of a hexagon.

First, find the sum of the interior angles of a hexagon:

$$(6 - 2) \times 180° = 720°$$

You can now write the expression and solve:

$$x + 114° + 90° + 147° + 125° + 143° = 619°$$

$$x + 619° = 720°$$

$$x + 619 \, (- 619) = 720 \, (- 619)$$

$$x = 101°$$

The measure of the angle labeled x is 101°.

HELPFUL POINT

In some instances, the value of the variable will not always represent the angle measure. You may need to substitute the value of x into the expression that represents the missing angle measure.

Let's take a look at another example. Suppose the interior angles of a quadrilateral are $3x°$, 50°, 40°, and 60°. Determine the value of the angle that measures $(3x)°$.

First, you need to write an equation. The sum of the angles of a quadrilateral is 360°, so you set up and solve as follows:

$$3x + 50° + 40° + 60° = 360°$$

$$3x + 150° = 360°$$

$$3x = 210°$$

$$x = 70°$$

Now that you have the value of x, you can determine the value of $3x°$:

$$3 \times 70 = 210°$$

The value of the fourth angle measure is 210°.

The Sum of the Exterior Angles of a Convex Polygon

An exterior angle of a polygon is the angle that forms a linear pair with an interior angle. It is the angle between any side of the polygon and a line extended from the next side. For example, in Figure 11.9, the angles labeled 1, 2, 3, 4, and 5 each represent exterior angles.

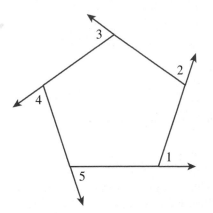

Figure 11.9: *The exterior angles of a regular pentagon.*

If you know the measure of an interior angle, you can find the measure of its exterior angle. Because Figure 11.9 is a regular pentagon, the measure of each of its angles is as follows:

$$\frac{(n - 2) \times 180°}{n} = \frac{(5 - 2) \times 180°}{5} = \frac{540}{5} = 108$$

∠1 forms a linear pair with an interior angle measuring 108°. You can write and solve the following equation to find the measure of ∠1; let the measure of ∠1 equal x:

$$x + 108° = 180°$$

$$x = 72°$$

So ∠1 is 72°. Because the pentagon is regular, you now know that the other exterior angles equal 72°, too.

Now that you know what exterior angles are, let's investigate the sum of exterior angles. Take a look at Figure 11.10A.

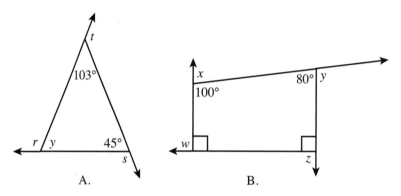

Figure 11.10: *Finding the sum of the exterior angles of a triangle (A) and quadrilateral (B).*

You can find the measure of ∠y with the Triangle Sum Theorem:

$$103° + 45° + y = 180°$$

$$148° + y = 180°$$

$$y = 32°$$

The measure of ∠r is determined by definition of a linear pair (see Chapter 4):

$$r + y = 180°$$

$$r + 32° = 180°$$

$$r + 32° (- 32°) = 180° (- 32°)$$

$$r = 148°$$

So $\angle r$ is 148°. Because $103° + 45° + y = 180°$ and $r + y = 180°$, you can also find r using the substitution property of equality (see Chapter 6):

$103° + 45° + y = r + y$

$103° + 45° = r$

$148° = r$

HELPFUL POINT

The exterior angle of a triangle also equals the sum of the nonadjacent interior angles.

To find the measures of $\angle s$ and $\angle t$, just plug in as you did before:

$s + 45° = 180°$

$s + 45° (-45°) = 180° (-45°)$

$s = 135°$

$t + 103° = 180°$

$t + 103° (-103°) = 180° (-103°)$

$t = 77°$

Finally, you add the angles together to determine the sum of the exterior angles:

$r + s + t =$ sum of the exterior angles

$148° + 135° + 77° = 360°$

The sum of the exterior angles of the triangle is 360°.

Let's look at one more example using the quadrilateral in Figure 11.10B. First, determine the measure of $\angle w$, $\angle x$, $\angle y$, and $\angle z$ using the definition of a linear pair:

$$w + 90° = 180°$$

$$w + 90° (- 90°) = 180° (- 90°)$$

$$w = 90°$$

$$x + 100° = 180°$$

$$x + 100° (- 100°) = 180° (- 100°)$$

$$x = 80°$$

$$y + 80° = 180°$$

$$y + 80° (- 80°) = 180° (- 80°)$$

$$y = 100°$$

$$z + 90° = 180°$$

$$z + 90° (- 90°) = 180° (- 90°)$$

$$z = 90°$$

Now add the angles together to get the sum of the exterior angles:

$$w + x + y + z = \text{sum of the exterior angles}$$

$$90 + 80 + 100 + 90 = 360°$$

Notice how the sum of the exterior angles, although a quadrilateral this time, is once again 360°. Regardless of the number of sides, the sum of the exterior angles of a convex polygon will always be 360°.

Perimeter and Area of Polygons

The distance around a two-dimensional figure is known as the *perimeter* of the figure. Think of it as the distance of a line that wraps around the figure. Perimeter is a one-dimensional measure, with units such as centimeters, inches, feet, kilometers, yards, and miles.

The number of square units that cover a two-dimensional figure represents the *area* of a figure. Area is a two-dimensional measure, with square units, such as cm^2, in^2, ft^2, yd^2, and mi^2.

Let's take a closer look at each of these concepts.

Perimeter

As you know, perimeter is the distance around a shape. If you think of it in the context of a backyard, the length of fencing needed to enclose the backyard represents the perimeter.

Let's work through an example, given figures on a coordinate grid, as shown in Figure 11.11. If you look at shape A, the length of the shape is 5 units and the width is 3 units; therefore, the perimeter is 5 + 3 + 5 + 3 = 16 units.

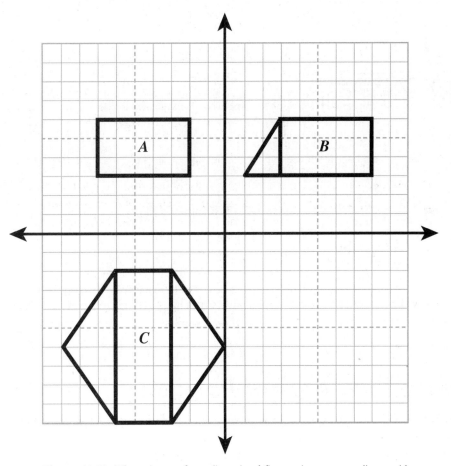

Figure 11.11: *The perimeter of two-dimensional figures given on a coordinate grid.*

Now take a look at shape B in Figure 11.11. The top length of the quadrilateral is 5 units, the bottom length is 7 units, and the height is 3 units. The length of the slanted side is determined using the Pythagorean Theorem (see Chapter 9):

$$a^2 + b^2 = c^2$$

$$3^2 + 2^2 = c^2$$

$$9 + 4 = c^2$$

$$13 = c^2$$

$$\sqrt{13} = c$$

Therefore, the perimeter of the quadrilateral is as follows:

$$5 + 7 + 3 + \sqrt{13} = 15 + \sqrt{13} \text{ units}$$

IT DOESN'T ADD UP

A common error is to include the lengths of line segments inside the two-dimensional figure. The perimeter is sum of the measures of the line segments on the outside only.

Now that you have a handle on the concept of perimeter, let's try a couple algebraic examples.

If you know the perimeter of a regular polygon, you can determine the length of one of its sides by dividing the perimeter by the number of sides. For example, if the perimeter of a square is 28 inches, the length of one of its sides can be determined by dividing 28 by 4:

$$28 \div 4 = 7 \text{ inches}$$

Suppose the perimeter of a regular hexagon is 126 ft. The measure of each of its side lengths must be equal. Therefore, take the perimeter and divide it by 6:

$$126 \div 6 = 21 \text{ ft}$$

Each side of the regular hexagon measures 21 ft.

Let's take a look at an irregular polygon. Suppose the length of a rectangle is four more than three times the width. Given the perimeter, you can write an equation and solve.

The perimeter of this rectangle is 40 cm. Let's determine the length and width. Let w equal width. The length equals $3w + 4$. Therefore, the equation to find the missing measures, given the perimeter, is as follows:

$$w + 3w + 4 + w + 3w + 4 = 40$$

$$8w + 8 = 40$$

$$8w + 8\ (-8) = 40\ (-8)$$

$$8w = 32$$

$$8w\ (\div 8) = 32\ (\div 8)$$

$$w = 4$$

The width is 4 cm, so plug it into the equation to get the length:

$$3(4) + 4 = 16 \text{ cm}$$

You can check your answer by adding the widths and lengths of the rectangle:

$$4 + 16 + 4 + 16 = 40 \text{ cm}$$

Area

As you know, area is the number of square units that covers the surface. Referring back to the backyard example, the number of square units of grass needed to cover the backyard represents the area.

Take a look at Figure 11.11 again. The number of square units that cover each of these shapes is the area. In shape A, it is easy enough to count the number of square units. Fifteen square units cover the shape; therefore, the area of the shape is 15 units^2.

HELPFUL POINT

Think of a tiled kitchen floor. The number of 1×1-foot tiles that cover the kitchen floor would be the area of the floor in square feet.

The formula for the area of the rectangle can be determined from the formula for the area of a triangle. The area of a triangle is $\frac{1}{2}bh$. Because the rectangle is made up of two triangles, the area of the rectangle is $2(\frac{1}{2}bh)$, or simply bh. In shape A, the base length is 5 units and the height is 3 units. Therefore, the area of shape A is the following:

$$5 \times 3 = 15 \text{ units}^2$$

Now let's find the area of shape B in Figure 11.11. In this case, you can't simply count square units. Because the shape is made up of a rectangle and a triangle, you can find the area of each shape and add it together. The area of the rectangle is as follows:

$$5 \times 3 = 15 \text{ units}^2$$

The area of the triangle is the following:

$$\frac{1}{2}(2 \times 3) = 3 \text{ units}^2$$

Therefore, the area of the quadrilateral is the area of the rectangle plus the area of the triangle:

$$15 + 3 = 18 \text{ units}^2$$

Shape C in Figure 11.11 is a hexagon. If you break it up into two triangles and one rectangle, it is easy to find the area. The length of the top and bottom of the hexagon is 3 units, while the height is 8 units; therefore, the area of the rectangle in the hexagon is 24 units^2:

$$3 \times 8 = 24 \text{ units}^2$$

The area of the two triangles is twice the area of one of the triangles:

$$2(\frac{1}{2}[8\times3]) = 8 \times 3 = 24 \text{ units}^2$$

The area of the hexagon is the area of the triangles and rectangle added together:

$$24 + 24 = 48 \text{ units}^2$$

You can find the area of any regular *n*-gon by dividing it into congruent triangles. For example, in Figure 11.12, the length of one side of the regular hexagon is 6 cm. Given the length from the center to a vertex point, you can determine the height of one of these triangles using the Pythagorean Theorem.

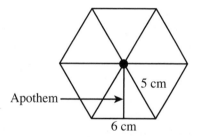

Figure 11.12: *Using the apothem to find the area of a regular polygon.*

The base length of the right triangle is half of 6 cm, or 3 cm. The hypotenuse is 5 cm. The height *a* of the triangle, also known as the *apothem*, is determined by writing and solving the following equation:

$$a^2 + 3^2 = 5^2$$

$$a^2 + 9 = 25$$

$$a^2 = 16$$

$$a = 4 \text{ cm}$$

The area of the triangles is half the length of the base times the apothem:

$$\tfrac{1}{2}(6 \times -4) = 12 \text{ cm}^2$$

There are six congruent triangles in the regular hexagon. Therefore, the area of the hexagon is as follows:

$$6 \times 12 = 72 \text{ cm}^2$$

HELPFUL POINT

You can also determine the area of the hexagon by taking half of the product of the apothem and the perimeter of the hexagon:

$$\tfrac{1}{2}(4 \times 6 \times 6) = 72 \text{ cm}^2$$

In general, the area of any regular *n*-gon is half the length of the apothem times the perimeter of the *n-gon*.

The Least You Need to Know

- Polygons can be more specifically classified based on the number of sides, side measures, and angle measures.
- The sum of the interior angles is always determined by multiplying two less than the number of sides by 180.
- The sum of the exterior angles of any n-gon is always 360°.
- The area of a polygon can be determined by breaking it up into familiar figures, such as triangles and rectangles, and finding the areas for those.

Special Quadrilaterals

In the previous chapter, you learned that quadrilaterals are four-sided polygons. Think about a piece of paper or a kite; not only are these representations of quadrilaterals, but they also represent special quadrilaterals.

In this chapter, you study special quadrilaterals. We begin by defining for you special quadrilaterals based on the relationships among the sides, angles, diagonals, and lines of symmetry. Thereafter, we help you explore the relationships among each of the quadrilaterals.

The Family of Special Quadrilaterals

First, what exactly makes quadrilaterals "special"? It has to do with their features; for example, all of them have at least one pair of sides that are either parallel or congruent. To better understand the relationships among quadrilaterals, check out Figure 12.1.

In This Chapter

- Identifying, describing, and naming special quadrilaterals
- Using coordinate geometry to specify the quadrilateral
- Doing proofs for parallelograms, rectangles, rhombuses, squares, trapezoids, and kites

HELPFUL POINT

If you are faced with an always, sometimes, or never question relating one special type of quadrilateral to another, a flow chart can be used to determine your response. For example, a square is always a rectangle because it is connected to the rectangle and below it. However, a rectangle is sometimes a square because it is connected to the square yet above it.

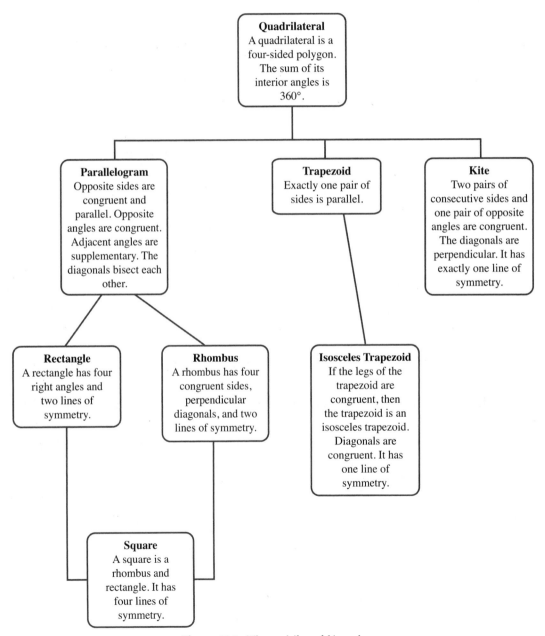

Figure 12.1: *The quadrilateral hierarchy.*

At the top of the flow chart, you have the quadrilateral. This is broken down into three different types: parallelogram, trapezoid, and kite. The parallelogram is further classified as a rhombus or rectangle. If the parallelogram is both a rhombus and a parallelogram, it is classified as a square. A trapezoid can be further classified as an isosceles trapezoid.

Parallelograms

A parallelogram is a quadrilateral with both pairs of opposite sides parallel. If a quadrilateral is a parallelogram, it is also true that the opposite sides are congruent, the opposite angles are congruent, the consecutive angles are supplementary, and the diagonals bisect each other.

Parallelograms in Detail

Figure 12.2 shows you examples of three different parallelograms, which we'll take a look at in detail.

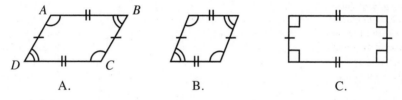

A.   B.   C.

Figure 12.2: *Some examples of parallelograms.*

In Chapter 7, we discussed special angle pair relationships when two parallel lines are crossed by a transversal. Using the labeled Figure 12.2A, extend $\overrightarrow{AD}$ so it appears to be a line. Do you see how $\overleftrightarrow{AB}$ and $\overleftrightarrow{DC}$ are parallel lines and $\overleftrightarrow{AD}$ is the transversal? Therefore, $\angle A$ and $\angle D$ are same-side interior angles. Same-side interior angles are supplementary and, as you can see, four pairs of adjacent angles are supplementary in a parallelogram. Figure 12.2B is more specifically a *rhombus*, while Figure 12.2C is more specifically a *rectangle*. What is the name of the quadrilateral that is both a rhombus and rectangle? That's right, a square!

DEFINITION

A **rhombus** is a parallelogram with four congruent sides, perpendicular diagonals, and two lines of symmetry. A **rectangle** is a parallelogram with four right angles, two lines of symmetry, and congruent diagonals.

Parallelogram Measurement Equations

You can use the properties of parallelograms to find missing measures. For example, suppose the measure of $\angle B$ in Figure 12.2A is 38°. Because $\angle B$ and $\angle C$ are supplementary, you can write and solve the following equation:

$$38° + m\angle C = 180°$$

$$38° (-38°) + m\angle C = 180°(-38°)$$

$$m\angle C = 142°$$

Now suppose the side lengths of parallelogram were given as algebraic expressions. Because opposite sides of parallelograms are congruent, you can write and solve an equation. Thereafter, you can substitute the value of the variable into each of the algebraic examples.

Using Figure 12.2A again, suppose $AB = 2x + 1$, $BC = y + 3$, $CD = 4x - 5$, and $DA = 4y$. Because $\overline{AB}$ is congruent to $\overline{CD}$, you can write and solve the following equation:

$$2x + 1 = 4x - 5$$

$$2x (-2x) + 1 (+5) = 4x (-2x) - 5 (+5)$$

$$6 = 2x$$

$$6 \div 2 = 2x \div 2$$

$$3 = x$$

Plugging the answer into the algebraic expressions, you get the following:

$$AB = 2(3) + 1 = 7 \text{ units}$$

$$CD = 4(3) - 5 = 7 \text{ units}$$

Notice how $AB = CD$, which is true of parallelograms. You can use this to justify that you solved the equation correctly.

Now solve for y and substitute to find BC and DA.

$$y + 3 = 4y$$

$$3 = 3y$$

$$1 = y$$

Again, you just plug the answer into the expressions:

$BC = 1 + 3 = 4$ units

$DA = 4(1) = 4$ units

You can see yet again that opposite segments are congruent, so you know you have the right answer.

Parallelogram Dimension Equations

For expressions that represent the dimensions of a parallelogram, you also integrate perimeter to determine the value of x. For example, suppose parallelogram $ABCD$ has dimensions $3y - 1$ and $y + 4$. If the perimeter of the parallelogram is 46 centimeters (cm), you can determine the value of y by writing and solving an equation. Because opposite sides of a parallelogram are congruent, you can take the sum of twice each of the dimension and set it equal to 46:

$2(3y - 1) + 2(y + 4) = 46$

$6y - 2 + 2y + 8 = 46$

$8y + 6 = 46$

$8y = 40$

$y = 5$

Because $y = 5$, the dimensions of the parallelogram are as follows:

$3(5) - 1 = 14$ cm

$5 + 4 = 9$ cm

HELPFUL POINT

It is always important to check your answer. For the dimension problem, you'd do so as follows:

$14 + 14 + 9 + 9 = 46$ cm

Now you know you solved the equation correctly!

Trapezoids

Unlike parallelograms, trapezoids have exactly one pair of parallel sides, which are the bases of the trapezoid.

Trapezoids in Detail

If you extend $\overline{XW}$ in trapezoid $XYZW$ in Figure 12.3A, you will see that $\angle X$ and $\angle W$ are same-side interior angles, which are supplementary. If you extend $\overline{YZ}$, you'll see that $\angle Y$ and $\angle Z$ are also supplementary.

While a parallelogram has four pairs of adjacent angles supplementary, a trapezoid only has two.

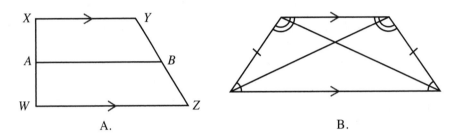

A.

B.

Figure 12.3: *Examples of trapezoids.*

The midsegment of a trapezoid is parallel to each base. Its measure is also one half of the sum of the measures of the bases. This is known as the *Midsegment Theorem for Trapezoids.* For example, in Figure 12.3A, A is the midpoint of $\overline{XW}$ and B is the midpoint of $\overline{ZY}$. Therefore, $AB = \frac{1}{2}(XY + WZ)$.

Figure 12.3B is more specifically an *isosceles trapezoid* because the sides other than the bases are also congruent. As seen in the figure, an isosceles trapezoid has two pairs of congruent angles, the diagonals are congruent, and it has one line of symmetry.

Trapezoid Measurement Equations

Like you did for parallelograms, you can use the properties of a trapezoid to find missing angle and length measures. For example, in Figure 12.4, you are given an isosceles trapezoid where P is the midpoint of $\overline{AC}$ and Q is the midpoint of $\overline{BD}$. Let's find the missing angles and the length of PQ.

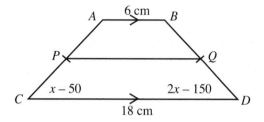

Figure 12.4: *Finding the missing measures of an isosceles trapezoid.*

Because you have an isosceles trapezoid, you know that the measure of $\angle C$ is equal to the measure of $\angle D$. Therefore, you can write and solve the following equation:

$x - 50 = 2x - 150$

$100 = 1x$

$100 = x$

Now that you know the value for x, you can determine the angle measures:

$m\angle C = x - 50$

$m\angle C = 100 - 50$

$m\angle C = 50°$

$m\angle D = 2x - 150 = 2(100) - 150$

$m\angle D = 2(100) - 150$

$m\angle D = 50°$

Because $\angle A$ is supplementary to $\angle C$, which is congruent to $\angle B$ and $\angle D$ respectively, you can find the measures of $\angle A$ and $\angle B$ by subtracting $m\angle C$ or $m\angle D$ from 180°:

$m\angle A$ and $B = 180 - 50$

$m\angle A$ and $B = 130°$

Lastly, you can determine *PQ* using the Midpoint Segment Theorem:

$$PQ = \frac{1}{2}(AB + DC).$$

$$PQ = \frac{1}{2}(6 + 18)$$

$$PQ = \frac{1}{2}(24)$$

$$PQ = 12 \text{ cm}$$

Kites

A kite is a quadrilateral that has two pairs of consecutive congruent sides; however, the opposite sides of a kite are not congruent.

Kites in Detail

If a quadrilateral is a kite, its diagonals are perpendicular and it has exactly one pair of congruent opposite angles. You can see this illustrated in Figure 12.5.

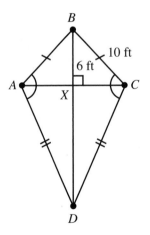

Figure 12.5: *Properties of a kite.*

Kite Measurement Equations

The properties of kites can help you solve for the missing lengths in one. For example, in Figure 12.5, you can find AB and XC. Because $\overline{AB}$ is congruent to $\overline{BC}$, you know AB must be 10 feet (ft).

Next, using the fact that a kite's diagonals are perpendicular, you can conclude $\triangle BXC$ is a right triangle. Therefore, you can use the Pythagorean Theorem to solve for the missing side:

$$(BX)^2 + (XC)^2 = (BC)^2$$

$$6^2 + (XC)^2 = 10^2$$

$$36 + (XC)^2 = 100$$

$$(XC)^2 = 64$$

$$XC = 8 \text{ ft}$$

Proving Parallelograms

The best way to understand the properties of quadrilaterals is to apply them. In geometry, these properties are integrated in coordinate geometry and proofs. Let's take a closer look at how to do this for parallelograms.

Parallelograms on a Coordinate Grid

When a four-sided figure is on a coordinate plane, you can use the distance formula and the slope formula to determine if sides are congruent and parallel, respectively (see Chapter 4), both of which apply to parallelograms.

> **HELPFUL POINT**
>
> Suppose you are given three coordinates of a quadrilateral. You can use the properties of the specified quadrilateral to find the fourth ordered pair.

Take a look at quadrilateral *ABCD* in Figure 12.6.

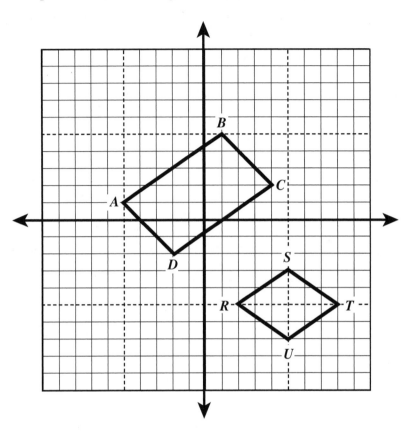

Figure 12.6: *Parallelograms on a coordinate plane.*

Remember, for the distance formula, XY with ordered pairs $X(x_1, y_1)$ and $Y(x_2, y_2)$ is determined by $\sqrt{(x_1 - x_2)^2 + (y_1 - y_2)^2}$. The ordered pairs of quadrilateral *ABCD* are *A* (-5, 1), *B* (1, 5), *C* (4, 2), and *D* (-2, -2); therefore, they are set up and solved as follows:

$$AB = \sqrt{(-5-1)^2 + (1-5)^2} = \sqrt{52} = 2\sqrt{13}$$

$$BC = \sqrt{(1-4)^2 + (5-2)^2} = \sqrt{18} = 3\sqrt{2}$$

$$CD = \sqrt{(4--2)^2 + (2--2)^2} = \sqrt{52} = 2\sqrt{13}$$

$$DA = \sqrt{(-2--5)^2 + (-2-1)^2} = \sqrt{18} = 3\sqrt{2}$$

Because *AB* = *CD* and *BC* = *DA,* you know the opposite sides have the same length.

Next, let's prove opposite sides are parallel. Remember, the slope of $\overline{XY}$ with ordered pairs $X\,(x_1, y_1)$ and $Y\,(x_2, y_2)$ is determined by $\dfrac{y_1 - y_2}{x_1 - x_2}$. So you set up and solve as follows:

$$\text{Slope of } \overline{AB} = \frac{1-5}{-5-1} = \frac{-4}{-6} = \frac{2}{3}$$

$$\text{Slope of } \overline{CD} = \frac{2--2}{4--2} = \frac{4}{6} = \frac{2}{3}$$

$$\text{Slope of } \overline{BC} = \frac{5-2}{1-4} = -\frac{3}{3} = -1$$

$$\text{Slope of } \overline{DA} = \frac{-2-1}{-2--5} = -\frac{3}{3} = -1$$

Because the opposite sides have the same slope, they are parallel. You have proved that opposite sides are congruent and parallel, which means quadrilateral $ABCD$ is a parallelogram.

Let's follow the same procedure for quadrilateral $RSTU$ in Figure 12.6. You first determine the measure of each side using the distance formula:

$$RS = \sqrt{(2-5)^2 + (-5--3)^2} = \sqrt{13}$$

$$ST = \sqrt{(5-8)^2 + (-3--5)^2} = \sqrt{13}$$

$$TU = \sqrt{(8-5)^2 + (-5--7)^2} = \sqrt{13}$$

$$UR = \sqrt{(5-2)^2 + (-7--5)^2} = \sqrt{13}$$

Not only are the opposite sides congruent, but all four sides are congruent! Next, you determine the slope of each line segment with the slope formula:

$$\text{Slope of } \overline{RS} = \frac{-5--3}{2-5} = \frac{2}{3}$$

$$\text{Slope of } \overline{UT} = \frac{-7--5}{5-8} = \frac{2}{3}$$

$$\text{Slope of } \overline{RU} = \frac{-5--7}{2-5} = -\frac{2}{3}$$

$$\text{Slope of } \overline{ST} = \frac{-3--5}{5-8} = -\frac{2}{3}$$

This shows the opposite sides have the same slope. Based on this information, you can conclude that quadrilateral *RSTU* is a rhombus.

If you recall, a rhombus also has perpendicular diagonals:

$$\text{Slope of } \overline{RT} = \frac{-5--5}{2-8} = -\frac{0}{4} \text{ (zero slope)}$$

$$\text{Slope of } \overline{SU} = \frac{-3--7}{5-5} = \frac{4}{0} \text{ (undefined slope)}$$

> **HELPFUL POINT**
>
> When the slope of a line is zero, the line is horizontal. When the slope of a line is undefined, the line is vertical.

Therefore, instead of determining if all four sides have the same length, you could have determined the relationship of the diagonals.

Parallelogram Two-Column Proofs

You can also use the angle pair relationships you learned in Chapter 7 to prove a quadrilateral is a parallelogram. For example, take a look at Figure 12.7.

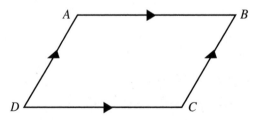

Figure 12.7: *How do you prove this is a parallelogram?*

Given $AB \parallel DC$, $AD \parallel BC$, and $m \angle A = m \angle C$, you prove quadrilateral *ABCD* is a parallelogram by proving $\angle B$ and $\angle D$ are also congruent. You can use the following as a guideline for any parallelogram proofs.

Statement	Reason
1. $AB \parallel DC$, $AD \parallel BC$, $m\angle A = m\angle C$	Given
2. $m\angle A + m\angle D = 180°$ $m\angle B + m\angle C = 180°$	Consecutive Interior Angles Theorem
3. $m\angle A + m\angle D = 180°$ $m\angle B + m\angle A = 180°$	Substitution property of equality
4. $m\angle A + m\angle D = m\angle B + m\angle A$	Substitution property of equality
5. $m\angle D = m\angle B$	Subtraction property of equality
6. $\angle D \cong \angle B$	Definition of congruent angles
7. $ABCD$ is a parallelogram	Definition of a parallelogram

By proving both pairs of opposite angles are congruent, you proved the quadrilateral to be a parallelogram.

Proving Trapezoids and Kites

Like you did with parallelograms, you can use the distance formulas you learned in Chapter 4 to determine whether quadrilaterals are trapezoids or kites. (Refer back to the characteristics shown in the quadrilateral flow chart if you need a reminder of the properties of each type.)

Using Properties of Trapezoids

The following are the two conditions that must be met for a trapezoid to be classified more specifically as an isosceles trapezoid:

1. If a trapezoid has a pair of congruent base angles, it is an isosceles trapezoid.

2. If the diagonals of the trapezoid are congruent, it is an isosceles trapezoid.

Let's try out an example. Quadrilateral $XYZW$ has vertices of X (4, 5), Y (-3, 3), Z (-6, -13), and W (6, -2). To determine if these are the vertices of an isosceles trapezoid, find out the length of its diagonals using the distance formula:

$$XZ = \sqrt{(4--6)^2 + (5--13)^2} = \sqrt{424} = 2\sqrt{106}$$

$$YW = \sqrt{(-3-6)^2 + (3--2)^2} = \sqrt{106}$$

Because the diagonals are not congruent, $XYZW$ is not an isosceles trapezoid.

Using Properties of Kites

If a quadrilateral is a kite, two pairs of consecutive sides are congruent and the diagonals are perpendicular.

For example, you can use this information to determine whether the points A (-2, 0), B (0, 2), C (5, 0), and D (0, -2) are the vertices of a kite by first plugging them into the distance formula:

$$AB = \sqrt{(-2-0)^2 + (0-2)^2} = \sqrt{8} = 2\sqrt{2}$$

$$AD = \sqrt{(-2-0)^2 + (0--2)^2} = \sqrt{8} = 2\sqrt{2}$$

$$CB = \sqrt{(5-0)^2 + (0-2)^2} = \sqrt{29}$$

$$CD = \sqrt{(5-0)^2 + (0--2)^2} = \sqrt{29}$$

So you know two pairs of consecutives sides are congruent. Next, determine the slope of $\overrightarrow{AC}$ and $\overrightarrow{BD}$:

$$\text{Slope of } \overrightarrow{AC} = \frac{0-0}{-2-5} = -\frac{0}{7}$$

$$\text{Slope of } \overrightarrow{BD} = \frac{2--2}{0-0} = \frac{4}{0}$$

The slope of $\overrightarrow{AC}$ is 0, which is the slope of a horizontal line, while the slope of $\overrightarrow{BD}$ is undefined, which is the slope of a vertical line. A horizontal line is always perpendicular to a vertical line; therefore, the diagonals are perpendicular. This means quadrilateral $ABCD$ is a kite.

The Least You Need to Know

- There are three types of special quadrilaterals: parallelograms, trapezoids, and kites.
- Each quadrilateral has a special equation associated with it to help you find missing measurements.
- Writing proofs using angle pair relationships can help you prove whether a quadrilateral is a parallelogram.
- You can use the distance and slope formulas to determine the type of quadrilateral.

CHAPTER 13

Similar Figures

When you studied the basics of geometry in your prealgebra class, you may have defined two figures as being congruent if they are the same size and same shape. You may have also defined similar figures as having the same shape but a different size. In this chapter, we help you refine this definition by exploring the conditions necessary as it relates to side lengths and angle measures.

Ratios and Proportions

A *ratio* compares two quantities. A *proportion* is a type of equation that's expressed when you set two ratios equal to one another.

In This Chapter

- Defining ratios and proportions
- Exploring dilations on a coordinate plane
- Indirectly measuring objects with similar triangles
- Using similarity theorems to prove two triangles are similar

> **DEFINITION**
>
> A **ratio** expresses the relationship between two quantities a and b written as a to b, $a:b$, or $\dfrac{a}{b}$. A **proportion** is an equation written as two ratios equal to each other.

For example, suppose you are making a batch of cookies for 12 friends but have a recipe that serves 18 people. How will you adjust the measurements in the recipe to serve 12 people? Let's say you need 6 tablespoons of sugar when serving 18, so you need to know how much sugar you will use for 12.

To create a ratio for this, let t equal tablespoons and p equal the number of people. You can mathematically compare tablespoons to the number of people as $t:p$ or $\dfrac{t}{p}$. In your cookie problem, the recipe calls for 6 tablespoons:18 people.

Now, you need to determine the number of tablespoons of sugar needed to serve 12 people. In order to do that, write a proportion with x standing in for that unknown value:

$$\frac{tablespoons}{people} \qquad \frac{6}{18} = \frac{x}{12}$$

Finally, cross-multiply to solve:

$(6)(12) = 18x$

$72 = 18x$

$4 = x$

As you can see, you need 4 tablespoons of sugar to make a batch of cookies for 12 people.

Another way to solve the proportion is to simplify 6:18 to 1:3. When you cross-multiply, that would give you $3x = 12$, which when divided by 3, brings you to the same answer of $x = 4$.

Now that you have a basic understanding of ratios and proportions, let's apply these concepts to determine unknown measures of geometrical figures.

Similar Polygons

Figures are similar when their corresponding side lengths are proportional and their corresponding angles are equal. Unlike congruent figures, similar figures don't have equal side lengths. Instead, you can think of similar figures as enlargements or reductions.

In Figure 13.1, trapezoid $ABCD$ is similar to trapezoid $XYZW$.

HELPFUL POINT

The symbol for similar is ~ . For example, in Figure 13.1, trapezoid *ABCD* ~ trapezoid *XYZW*.

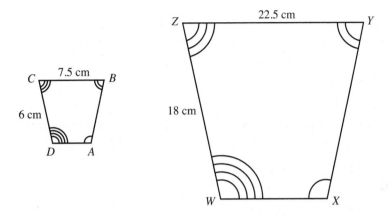

Figure 13.1: *Similar trapezoids.*

In the figure, the corresponding angles are congruent and the corresponding sides are proportional—that is, $\dfrac{BC}{CD} = \dfrac{YZ}{ZW}$. More specifically, the proportionality of these figures is validated by $\dfrac{7.5}{6} = \dfrac{21.5}{18}$. What do you notice? That's right, the ratios are equivalent:

$$7.5 \div 6 = 1.25$$

$$22.5 \div 18 = 1.25$$

The side lengths of trapezoid *XYZW* are three times the side lengths of trapezoid *ABCD*, or reverse that and say the smaller figure is one third the size of the larger figure. Let's use the latter to write a ratio for the dimensions of trapezoid *ABCD* to trapezoid *XYZW*, which is 1:3.

When all sides of one figure are multiplied by a number to find the lengths of corresponding sides of another figure, this number is known as the *scale factor*. The scale factor of trapezoid *ABCD* to trapezoid *XYZW* is 3, while the scale factor of trapezoid *XYZW* to trapezoid *ABCD* is one third. As you can see, this directly relates to the ratio between the two.

IT DOESN'T ADD UP

All congruent figures are similar; the scale factor for them is always 1. However, all similar figures are not always congruent, because the scale factor is not always 1.

A Numerical Example

If you know two polygons are similar, you can use the definition of similar figures to find missing measures. For example, in Figure 13.2, $\triangle ABC \sim \triangle XYZ$; you can determine AC, YZ, and the measure of $\angle Z$.

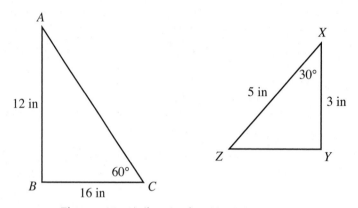

Figure 13.2: *Similar triangles with missing measures.*

The letters in the similarity statement correspond in the order given—for instance, $\angle A$ is congruent to $\angle X$, or $\overline{AC}$ is proportional to $\overline{XZ}$. Using this information, you can write and solve proportions to find AC and YZ:

$$\frac{XY}{XZ} = \frac{AB}{AC}$$

$$\frac{3}{5} = \frac{12}{AC}$$

$$60 = 3(AC)$$

$$60 \div 3 = 3AC \div 3$$

$$20 = AC$$

$$\frac{XY}{YZ} = \frac{AB}{BC}$$

$$\frac{3}{YZ} = \frac{12}{16}$$

$$48 = 12(YZ)$$

$$48 \div 12 = 12YZ \div 12$$

$$4 = YZ$$

So *AC* is 20 inches, and *YZ* is 4 inches.

Because $\angle Z$ is congruent to $\angle C$, the measure of $\angle Z$ is 60°.

> **HELPFUL POINT**
>
> You do not always have to use cross products to find the solution to a proportion. For example, in Figure 13.2, *AB* is 4 times longer than *XY*; therefore, *AC* is 4 times *XZ*.

An Algebraic Example

Now that you have a handle on how to write and solve a proportion to find missing measures of similar polygons, let's take a look at an algebraic application.

Suppose quadrilateral *BARK* is similar to quadrilateral *SLIP*. If *BA* = 6 meters, *KR* = 9 meters, *SL* = 16 meters, and *PI* = $5x - 1$ meters, write and solve a proportion to find the value of *x*:

$$\frac{BA}{KR} = \frac{SL}{PI}$$

$$\frac{6}{9} = \frac{16}{5x-1}$$

$$6(5x - 1) = 16 \times 9$$

$$30x - 6 = 144$$

$$30x = 150$$

$$30x \div 30 = 150 \div 30$$

$$x = 5$$

The value of *x* is 5. You can now substitute $x = 5$ into $5x - 1$ and check to see if your answer makes sense:

$$5(5) - 1 = 24$$

So *PI* is 24 meters. Both 6:9 and 16:24 reduce to 2:3; therefore, you know the value you found for *x* is correct.

Similar Polygons on a Coordinate Plane

If one figure is a dilated image of another, the figures are similar. A dilation occurs when a polygon is enlarged or reduced. To dilate a figure on the coordinate plane, you multiply the coordinates of all of its vertices by the scale factor.

For example, in Figure 13.3, hexagon *ABCDEF* has vertices with coordinates *A* (-1,3), *B* (2, 4), *C* (4,-1), *D* (3,-3), *E* (1,-3), and *F* (-3,0). If you want make the figure 1.5 times larger, you can dilate it by multiplying each of the coordinates by 1.5, as shown in the following; the ′ after the letter indicates that is the new coordinate for each of the vertices:

A (-1, 3) → A' (-1.5, 4.5) D (3, -3) → D' (4.5, -4.5)

B (2, 4) → B' (3, 6) E (1, -3) → E' (1.5, -4.5)

C (4, -1) → C' (6, -1.5) F (-3, 0) → F' (-4.5, 0)

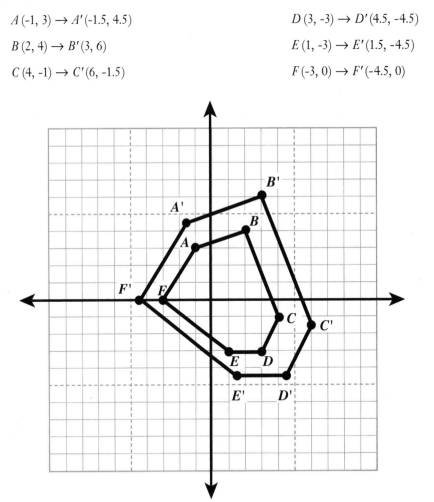

Figure 13.3: *Hexagon ABCDEF before and after dilation.*

The corresponding angles are congruent and the corresponding sides are proportional by a scale factor of 1.5; therefore, hexagon *ABCDEF* ~ hexagon *A′B′C′D′E′F′*.

Proportions with Area

Let's work with scale factor some more. If you multiply the dimensions of a figure by a scale factor, you affect its area.

Referring back to Figure 13.2, because triangle ABC is four times the size of triangle XYZ, the corresponding side lengths of the triangles are a ratio of 1:4. First, compare the areas of these triangles.

The area of $\triangle XYZ$ is the following:

$$\text{Area of } \triangle XYZ = \frac{1}{2}(YZ)(XY)$$

$$\frac{1}{2}(4)(3) = 6 \text{ square inches}$$

The area of $\triangle ABC$ is the following:

$$\text{Area of } \triangle ABC = \frac{1}{2}(BC)(AB)$$

$$\frac{1}{2}(16)(12) = 96 \text{ square inches}$$

Their areas compare in the ratio 6:96, which reduces to 1:16.

Let's explore how the ratio of proportionality compares to the ratio of the areas. The area of a triangle is half the base times the height. If you multiplied the base by 4 and the height by 4 to get the dimensions of the enlarged triangle, this means the area of the enlarged triangle is 4×4, or 16 times greater! This is illustrated as the following:

$$A_{ABC} = A_{XYZ} \times (\text{scale factor})^2$$

The proportional areas conjecture states if two corresponding sides of two similar polygons compare in ratio $a{:}b$, then their areas compare in ratio $a^2{:}b^2$. That is, 1:16 is $1^2{:}4^2$.

> **HELPFUL POINT**
>
> Area is a two-dimensional measure. Therefore, if you multiply the dimensions of a polygon by 4, the area will be 4×4 or 16 times greater.

Applications of Similar Triangles

Similar triangles have many different applications. You can use them to indirectly measure an object or to find missing measures. The following sections walk you through how to do both.

Indirect Measurement

Sometimes you can't measure objects directly with a tape measure or other tool because they're too tall. In these cases, you can use indirect measurement with similar triangles.

Suppose you wanted to find the height of a tree. To estimate the height of the tree, you can stand parallel to the tree, as shown in Figure 13.4.

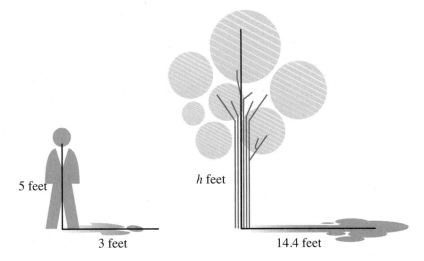

Figure 13.4: *Indirectly measuring the height of a tree.*

Say you are 5 feet tall and cast a shadow measuring 3 feet. The tree casts a shadow of 14.4 feet. Using the ratio height in feet:shadow in feet, you can determine the height of the tree by writing and solving the following proportion; let b = the height of the tree in feet:

$$\frac{5}{3} = \frac{b}{14.4}$$

$$3b = 72$$

$$3b \div 3 = 72 \div 3$$

$$b = 24 \text{ feet}$$

Therefore, the height of the tree is 24 feet.

Overlapping Triangles

If a line is drawn between two sides of a triangle and the line drawn is parallel to the third side, then similar triangles are created. If you look at Figure 13.5, $\triangle CAN$ is similar to $\triangle BAT$.

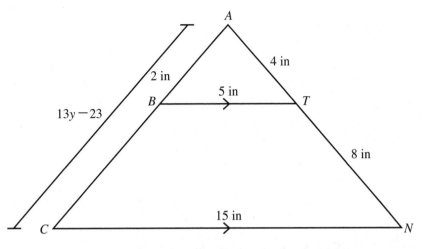

Figure 13.5: *Overlapping triangles.*

$\overline{BT}$ is parallel to $\overline{CN}$. By the Triangle Proportionality Theorem (see Chapter 8), if a line parallel to one side of a triangle intersects the other two sides, it divides the two sides proportionally; therefore, $\dfrac{AB}{BC} = \dfrac{AT}{TN}$.

By the Corresponding Angles Theorem, $\angle B$ is congruent to $\angle C$ and $\angle T$ is congruent to $\angle N$. By the reflexive property of congruence, $\angle A$ is congruent to $\angle A$. Therefore, in accordance with the definition of similar polygons, the corresponding angles are indeed congruent.

You can find the value of y by writing and solving a proportion, as follows:

> ☞ **HELPFUL POINT**
>
> When you are finding missing measures of similar triangles that overlap one another, separate the triangles so it is easier to interpret the information you are given and what you need to find.

$$\frac{AB}{BT} = \frac{AC}{CN} \qquad\qquad \frac{1}{3} = \frac{13y - 20}{15}$$

Because 15 is 3 times greater than 5, $13y - 23$ must be three times greater than 2:

$$13y - 20 = 6$$

$$13y = 26$$

$$y = 2$$

Therefore, AC measures $13(2) - 23 = 6$ inches. Because 6:15 reduces to 2:5, you know the answer is correct.

Now that you have a handle on the application of similar polygons in geometry, let's take a deeper look at similarity theorems that prove triangles are similar.

Proving Triangles Are Similar

If you recall from Chapter 10, you can use the SSS, SAS, ASA, or AAS Postulate to prove triangles are congruent. Like those, similar triangles have theorems to prove them. They are known as the AA Similarity Theorem, the SSS Similarity Theorem, and the SAS Similarity Theorem.

AA Similarity Theorem

The Angle-Angle (AA) Similarity Theorem states that if two angles of one triangle are congruent to two angles of another triangle, then the two triangles are similar. For example, in Figure 13.6, $\angle D$ is congruent to $\angle C$ and $\angle O$ is congruent to $\angle O$. Therefore, $\triangle DOG \sim \triangle CAT$.

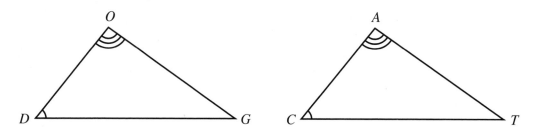

Figure 13.6: *These two triangles are similar according to the AA Similarity Theorem.*

Given Figure 13.7, you can write a proportionality statement based on the AA Similarity Theorem. In this case, let's find *AC*.

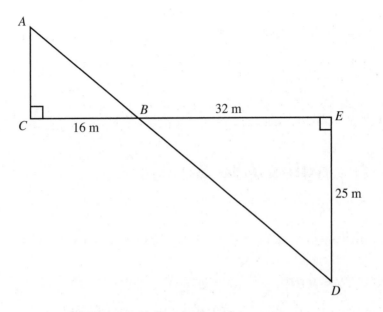

Figure 13.7: *Similar triangles with a common vertex.*

First of all, the measure of ∠*ABC* is congruent to ∠*DBE* by the Vertical Angles Congruency Theorem (see Chapter 10). You are also given that ∠*C* and ∠*E* are 90° each; therefore, two angles of Δ*ABC* are congruent to two angles of Δ*DBE*. So by the AA Similarity Theorem, Δ*ABC* is similar to Δ*DBE*.

Now that you have proved triangles to be similar, you can find $\overrightarrow{AC}$ by writing and solving the proportion, with *x* standing in for the missing measure:

$$\frac{x}{16} = \frac{25}{32}$$

$$32x = 25 \times 16$$

$$32x = 400$$

$$32x \div 32 = 400 \div 32$$

$$x = 12.5 \text{ meters}$$

So the measure of *AC* is 12.5 meters.

SSS Similarity Theorem

The Side-Side-Side (SSS) Similarity Theorem states that if the lengths of the corresponding sides of two triangles are proportional, then the triangles are similar. For example, Figure 13.8 has three triangles. Which triangles are similar?

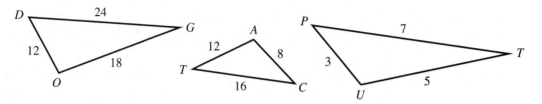

Figure 13.8: *Which triangles are similar according to the SSS Similarity Theorem?*

To decide which of these triangles are similar, you need to determine the ratios of the lengths of the corresponding sides.

The ratios of $\triangle DOG$ and $\triangle CAT$ are as follows:

$$\text{Ratio of the shortest sides} = \frac{DO}{CA} = \frac{12}{8} = \frac{3}{2}$$

$$\text{Ratio of the longest sides} = \frac{GD}{TC} = \frac{24}{16} = \frac{3}{2}$$

$$\text{Ratio of the other side} = \frac{OG}{AT} = \frac{18}{12} = \frac{3}{2}$$

Because the lengths of the corresponding sides of $\triangle DOG$ and $\triangle CAT$ are proportional, $\triangle DOG$ is similar to $\triangle CAT$.

Let's take a look at the ratios of the side lengths of $\triangle DOG$ and $\triangle PUT$:

$$\text{Ratio of the shortest sides} = \frac{DO}{PU} = \frac{12}{3} = \frac{4}{1}$$

$$\text{Ratio of the longest sides} = \frac{GD}{TP} = \frac{24}{7}$$

$$\text{Ratio of the other side} = \frac{OG}{UT} = \frac{18}{5}$$

Because the lengths of the corresponding sides of $\triangle DOG$ and $\triangle PUT$ are not proportional, $\triangle DOG$ is not similar to $\triangle PUT$.

SAS Similarity Theorem

The Side-Angle-Side (SAS) Similarity Theorem states that if an angle of one triangle is congruent to an angle of a second triangle and the lengths of the sides including these angles are proportional, then the triangles are similar.

If you refer back to Figure 13.5, you can prove that $\triangle BAT$ is similar to $\triangle CAN$ using the SAS Similarity Theorem. Begin by finding the ratios of the lengths of the corresponding sides to prove AC and AN are proportional to the lengths of the corresponding sides of $\triangle BAT$:

$$\frac{AC}{AB} = \frac{AB + BC}{AB} = \frac{2 + 4}{2} = \frac{6}{2} = \frac{3}{1}$$

$$\frac{AN}{AT} = \frac{AT + AN}{AT} = \frac{4 + 8}{4} = \frac{12}{4} = \frac{3}{1}$$

You also know that $\angle A$ is the included angle in both triangles. Therefore, by the SAS Similarity Theorem, $\triangle CAN$ is similar to $\triangle BAT$.

The Least You Need to Know

- A ratio compares two quantities. A proportion is a type of equation that's expressed when you set two ratios equal to one another.
- If one figure is a dilated image of another, the figures are similar.
- According to the Triangle Proportionality Theorem, if a line parallel to one side of a triangle intersects the other two sides, then it divides the two sides proportionally.
- The AA Similarity Theorem says that if two angles of one triangle are congruent to two angles of another triangle, then the triangles are similar.
- By the SSS Similarity Theorem, if the lengths of the corresponding sides of two triangles are proportions, then the triangles are similar.
- If an angle of one triangle is congruent to an angle of a second triangle and the lengths of the sides including these angles are proportional, then the triangles are similar, according to the SAS Similarity Theorem.

Three-Dimensional Figures

In this part, you relate geometry to the three-dimensional world you live in. You explore common three-dimensional figures and use two-dimensional concepts to explore measures of the three-dimensional figures. Whether needing to cover a three-dimensional figure or fill it, this part of the book helps you learn how.

CHAPTER 14

Solid Geometry

You live in a three-dimensional world. Take a look around you. Objects are not flat—they have length, width, and height. Even a sheet of paper has three dimensions. It may not appear that way at first, but if you stack papers on top of each other, you realize that the papers are three-dimensional.

Geometry is the study of size, shape, and position. Therefore, a book on geometry isn't complete without a look at the size, shape, and position of the three-dimensional objects in the world. In this chapter, we share information with you about common three-dimensional figures.

Solid Figures

Solid geometry is the study of geometry in three-dimensional space. The objects in this three-dimensional space are called *solid figures*. Solid figures have several properties, including faces, edges, and vertices (see Chapter 15 for information on surface area and volume). Faces are the surfaces on a solid figure, edges are line segments formed by the intersection of two faces, and vertices are the points formed by the intersection of three or more edges. Think of faces as planes, edges as lines, and vertices as points.

In This Chapter

* Classifying solid figures
* Relating three-dimensional figures to their two-dimensional representations with nets
* Using Euler's formula to determine the number of faces, edges, and vertices of a polyhedron

 **DEFINITION**

A **solid figure** is a three-dimensional figure.

The following solid figures are categorized into two groups: polyhedrons and nonpolyhedrons.

Polyhedrons

Some solid figures are referred to as *polyhedrons*. Polyhedrons are solid figures that consist of all-flat faces. Polyhedrons always have more edges than faces and vertices. Prisms and pyramids are two common types of polyhedrons.

A *prism* is a solid figure with *t* sides made up of polygons (see Chapter 11) and two parallel bases. But what is a base? A base is the bottom of a figure or the side that a figure stands on. So when looking at a picture of a solid like a prism, a base could either be on the bottom or facing out from a side. Check out Figure 14.1.

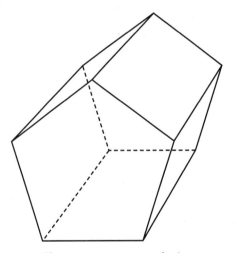

Figure 14.1: *A pentagonal prism.*

The prism in the figure is known as a pentagonal prism because the two parallel bases are pentagons (five-sided polygons). Notice how the prism is not standing on its bases. Because its faces are all flat, a prism can stand on any of its sides.

The faces, edges, and vertices of a prism vary based on the number of sides. For example, the rectangular prism in Figure 14.1 has 7 faces, 15 edges, and 10 vertices. A *triangular prism,* which you can see in Figure 14.2, has 5 faces, 9 edges, and 6 vertices. In the following figure, I have labeled a face, an edge, and a vertex.

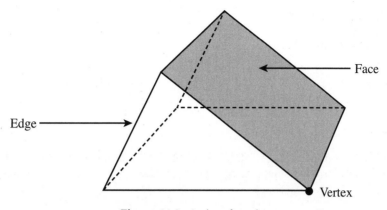

Figure 14.2: *A triangular prism.*

The prisms in Figures 14.1 and 14.2 are also known as *right prisms* because the bases are aligned one directly above the other. When the bases are not aligned one directly above the other, the prism is called an *oblique prism*. Figure 14.3 shows an example of an oblique prism.

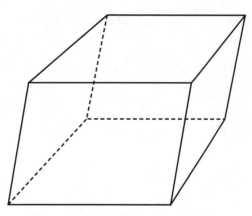

Figure 14.3: *An oblique rectangular prism.*

Pyramids are polyhedrons with sides made up of triangles and one polygon base. A pyramid is named by the shape of its base. For example, the pyramid in Figure 14.4 is called a *rectangular pyramid* because its base is a rectangle. A rectangular pyramid has 5 faces, 8 edges, and 5 vertices. As you can see, if the shape of the pyramid's base changes, the number of faces, edges, and vertices changes.

DEFINITION

A **pyramid** is a solid with a polygon base and triangular faces that meet at one vertex.

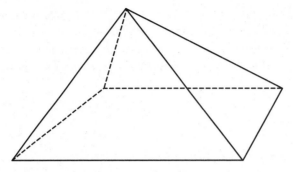

Figure 14.4: *A rectangular pyramid.*

The other faces of a pyramid, which are always triangles, meet at a single vertex called the *apex*. Because the apex is not directly above the center of the base in Figure 14.4, the pyramid can't be classified as a right pyramid; instead, it is referred to as oblique.

Nonpolyhedrons

Solid figures that don't have all-flat faces are nonpolyhedrons, or not polyhedrons. Cylinders, cones, and spheres are examples of nonpolyhedrons.

A *cylinder* is a solid figure with two parallel congruent circular bases and a curved surface. It is the curved surface that does not allow a cylinder to be classified as a polyhedron.

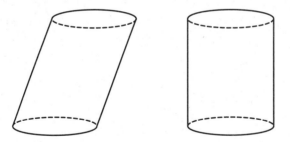

Figure 14.5: *Two cylinders—one oblique and one right.*

In Figure 14.5, the cylinder on the left is an *oblique cylinder*, because the two bases are not aligned directly above one another. The cylinder on the right is a *right cylinder*, because the two bases are directly aligned above one another. Cylinders have three faces (two bases and one curved surface), two edges that connect the curved surface to the bases, and no vertices.

A *cone* is another type of solid figure that is not a polyhedron because it has one circular base and a curved surface. Like prisms, pyramids, and cylinders, a cone can be either oblique or right. An *oblique cone* doesn't have an apex point directly above the center of the base, while a *right cone* has an apex point directly above the center of the base. Figure 14.5 shows both an oblique cone and a right cone.

DEFINITION

A **cone** is a solid with a circular base and one vertex.

Figure 14.6: *An oblique cone (left) and a right cone (right).*

Cones have two faces (the base and the curved surface), one edge that connects the base to the curved surface, and one vertex (called the *apex*).

A *sphere* is a round solid figure with every point on its surface equidistant from its center. Shaped like a ball, it has no faces, no edges, and no vertices. You can see an example of a sphere in Figure 14.7.

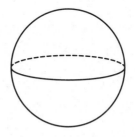

Figure 14.7: *A sphere.*

Nets

Every solid figure has a *net,* or two-dimensional representation. Think of a net as unfolding a solid figure and laying it out flat. Take a look at Figure 14.8.

A **net** is a two-dimensional representation of a three-dimensional figure.

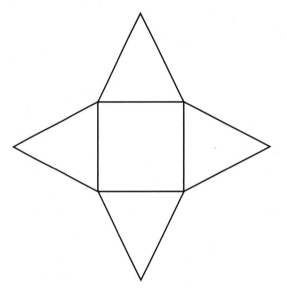

Figure 14.8: *A net of a square pyramid.*

The net in the figure is a square pyramid. Thinking of it as a three-dimensional object, the square is the base and the triangles fold up to meet at the apex.

Let's look at another example using a cube, or a prism whose faces are all squares. Because each face is a square, it is possible to draw several different nets of a cube. Figure 14.9 shows some potential options for the net of a cube. Which do you think will work?

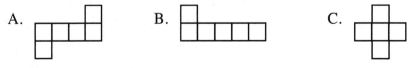

Figure 14.9: *Possible nets of a cube.*

Even though each of the figures from Figure 14.9 have six squares, that does not mean they will fold into a cube. *A* and *C* will fold into a cube, but *B* will not. If B is folded, there is no way to have both a top and bottom on the cube, because two of the faces will overlap.

Sometimes the net of a nonpolyhedron is harder to visualize. Let's think about a cylinder. After lifting the two bases up, the curved surface needs to be unrolled. When the curved surface is unrolled, it is a rectangle. Figure 14.10 shows the net of a cylinder.

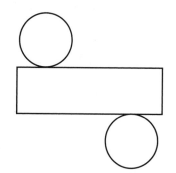

Figure 14.10: *The net of a cylinder.*

Euler's Formula

For every polyhedron, there is a relationship between the number of faces, edges, and vertices. Regardless of the polyhedron, the faces (F), edges (e), and vertices (V) can be plugged into Euler's formula. Euler's formula states that the sum of the vertices and faces minus the number of edges must equal 2. In equation form, the formula is the following:

$$V + F - E = 2$$

HELPFUL POINT

Euler's formula only applies to polyhedrons. Therefore, it does not work for cylinders, cones, and spheres.

Let's use a triangular prism to prove that Euler's formula is true. Previously, we told you a triangular prism has five faces, nine edges, and six vertices. Substitute these values into the formula to see if it is true:

$$V + F - E = 2$$

$$6 + 5 - 9 = 2$$

$$11 - 9 = 2$$

$$2 = 2$$

Because both sides of the equation are equal, the equation is true.

Euler's formula can also be used to find the number of faces, edges, or vertices without knowing the specific solid figure. For example, if we tell you that a polyhedron has four faces and four vertices, you can use Euler's formula to determine the number of edges:

$$V + F - E = 2$$

$$4 + 4 - E = 2$$

$$8 - E = 2$$

$$E = 6$$

So the polyhedron must have six edges. Once you know the number of faces, edges, and vertices, you can determine the type of polyhedron. In this case, a polyhedron with four faces, six edges, and four vertices is a triangular pyramid.

The Least You Need to Know

- Solid figures can be polyhedrons or nonpolyhedrons. Prisms and pyramids are examples of polyhedrons; cylinders, cones, and spheres are examples of nonpolyhedrons.

- Every solid figure has a net, or two-dimensional representation of the three-dimensional figure.
- Euler's formula represents a relationship between the number of faces, edges, and vertices in a polyhedron. You can use it to confirm the numbers for a polyhedron, even if you don't know the specific polyhedron.

CHAPTER 15

Measuring Solid Figures

Measuring solid figures is like putting together a present. The amount of candy you can put in the box is based on the volume of the box, while the paper you need is based on the surface area of the box. In this chapter, we look at the surface area and volume of solid figures. We also take you through how to identify similar solids, as well as dilation of solid figures.

Surface Area

All solid figures consist of faces and curved surfaces. Each face and curved surface has an area. The sum of these areas is the *surface area*, *S*, of the solid figure.

For example, a rectangular prism has six faces, each of which is a rectangle. If we find the area of each of these rectangles and add them together, we will have the surface area of the rectangular prism.

The **surface area** of a solid figure is the sum of the areas of the figure's faces and curved surfaces.

In the following sections, we give you the surface area formulas for different solid figures.

The Surface Area of Prisms and Cylinders

Prisms and cylinders both have two bases, so finding the surface area for each type isn't that different. In fact, the formulas for the surface area of a prism and the surface area of a cylinder are very similar.

First, let's focus on prisms. Prisms consist of sides called *faces.* Two of these faces are the bases of the prism. The other faces are called *lateral faces.* The area of the lateral faces is called the *lateral area, L,* which is part of what you need to get the surface area of a prism. Figure 15.1 shows a triangular prism; how do we find the lateral area?

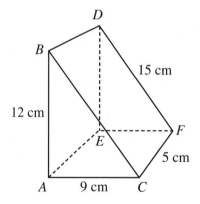

Figure 15.1: *A triangular prism.*

The two triangular faces are the bases of the prism, while the three rectangular faces are the lateral faces. Recall from Chapter 11 that the area of a rectangle is found by multiplying the length of the base and the length of the height ($A = bh$). To find the lateral area, L, you need to find the sum of the areas of the lateral faces:

Area of rectangle *ABDE* = bh = 5 × 12 = 60 square centimeters

Area of rectangle *ACFE* = bh = 9 × 5 = 45 cm²

Area of rectangle *BCFD* = bh = 15 × 5 = 75 cm²

L = 60 + 45 + 75 = 180 cm²

You can also find the lateral area of a prism by multiplying the perimeter of the base, P, by the height, h:

$$L = Ph$$

For example, the base of the prism in Figure 15.1 is a triangle with sides measuring 9 centimeters, 12 centimeters, and 15 centimeters. Therefore, the perimeter of the base is $9 + 12 + 15 = 36$ centimeters. The height of the triangular prism is 5 centimeters. Substitute into the formula to find the lateral area:

$$L = Ph$$

$$L = 36 \times 5$$

$$L = 180 \text{ cm}^2$$

As you can see, this formula gives you the same answer as the previous one—the lateral area of the triangular prism is 180 cm².

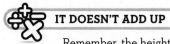 **IT DOESN'T ADD UP**

Remember, the height joins the two bases of the prism. Therefore, don't get tripped up and think the height is 12 cm for Figure 15.1; it's actually 5 cm.

Remember, surface area is the sum of the area of the faces and curved surfaces of a solid. Therefore, after finding the lateral area, you only need to find the area of the two bases. Figure 15.1 shows a triangular prism, so the bases are triangles. Therefore, you can substitute into the area of a triangle formula (see Chapter 11) to find the area of the base:

$$A = \frac{1}{2}bh$$

$$A = \frac{1}{2}(9)(12)$$

$$A = \frac{1}{2}(108)$$

$$A = 54$$

The area of the triangular base is 54 cm². Because there are two bases, you need to double that area to 108 cm². The area of the bases added to the lateral area is the surface area of the triangular prism:

$$180 + 108 = 288 \text{ cm}^2$$

So the formula for finding the surface area of a prism is as follows:

$S = L + 2B$ or $S = Ph + 2B$

In this formula, S stands for "surface area," while B stands for "area of the base."

Let's look at another example that uses this formula. Figure 15.2 shows a rectangular prism. How do you find the surface area?

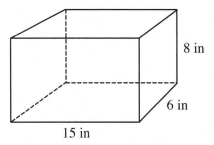

Figure 15.2: *A rectangular prism.*

> **IT DOESN'T ADD UP**
>
> The surface area formulas we show you in this chapter are for right solids. Therefore, don't try to use them for oblique solids, or you won't get the correct answer.

Because the solid in the figure is a rectangular prism, the perimeter of the base is found by multiplying 2 times the length and 2 times the width ($P = 2l + 2w$). The area of the base is found by multiplying the base and the height ($A = bh$). Here's how you solve for the surface area of a prism using these pieces:

$S = Ph + 2B$

$S = (2l + 2w)h + 2(bh)$

$S = ([2 \times 15] + [2 \times 6])8 + 2(15 \times 6)$

$S = (30 + 12)8 + 2(90)$

$S = (42)8 + 180$

$S = 336 + 180$

$S = 516 \text{ in}^2$

The surface area of the rectangular prism is 496 in².

Like for prisms, the surface area of a cylinder can be found using the formula $S = L + 2B$. However, finding the lateral area, L, involves some different steps, because the lateral area of a cylinder is the area of the curved surface.

When a cylinder is unrolled into its two-dimensional net, the curved surface is a rectangle. The height of the rectangle is the height of the cylinder, and the base of the rectangle is equal to the circumference of the base of the cylinder. Take a look at Figure 15.3.

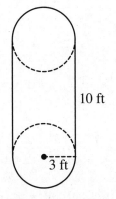

10 ft

3 ft

Figure 15.3: *A cylinder.*

To find the lateral area of the cylinder, you multiply the circumference of the base and the height of the cylinder. This is done as follows:

$L = 2\pi rh$

$L = 2\pi(3)(10)$

$L = 60\pi$

$L \approx 188.4$

Using $2\pi rh$ as the lateral area, the surface area formula of a cylinder can be written as follows:

$S = L + 2B$ or $S = 2\pi rh + 2B$

The base of a cylinder is a circle. Therefore, the area of the base, B, can be found using the formula for the area of a circle ($A = \pi r^2$). Using this information, here's how you find the surface area of the cylinder in Figure 15.3:

$$S = L + 2B$$
$$S = 2\pi rh + 2\pi r^2$$
$$S = 2\pi(3)(10) + 2\pi(3)^2$$
$$S = 60\pi + 18\pi$$
$$S = 78\pi$$
$$S = 244.92$$

The surface area of the cylinder is approximately 244.92 ft².

The Surface Area of Pyramids and Cones

Pyramids and cones have a similar formula for finding the surface area. Unlike prisms and cylinders, pyramids and cones only have one base. To reflect that, the surface area formula does not multiply the area of the base by two. The following is the formula for the surface area of pyramids and cones:

$$S = L + B$$

Let's take a look at the pyramid first. The base of a pyramid can have many different shapes; therefore, the formula for the area of the base of the pyramid depends on the type of pyramid given.

The lateral sides of a prism are always triangles. As you know, the area of a triangle is found by multiplying one half times the base and the height ($A = \frac{1}{2}bh$). The base length of the lateral sides is the same as the length of the sides of the base of the pyramid. However, the height of the lateral sides—known as the slant height, l—is not as easy to determine. To make this clearer, take a look at Figure 15.4.

The pyramid in the figure is a square pyramid, which means the base of the pyramid is a square. The height of the pyramid, $\overline{EF}$, is perpendicular to the base and intersects the base at the center; $\overline{EG}$ is the slant height of the pyramid.

$\overline{EF}, \overline{FG}$, and $\overline{EG}$ form a right triangle; therefore, the slant height can be found by using the Pythagorean Theorem (see Chapter 9).

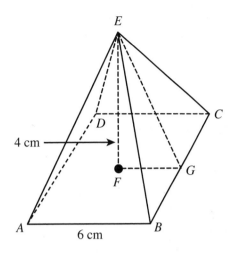

Figure 15.4: *A square pyramid.*

HELPFUL POINT

Because $\overline{EF}$ intersects the center of the base, the length of $\overline{FG}$ is equal to half of the length of $\overline{AB}$.

$$a^2 + b^2 = c^2$$
$$EF^2 + FG^2 = EG^2$$
$$4^2 + 3^2 = EG^2$$
$$16 + 9 = EG^2$$
$$25 = EG^2$$
$$\sqrt{25} = \sqrt{EG^2}$$
$$5 = EG$$

Now that you know that the slant height of the pyramid is 5 cm, you can find the lateral area, L. The formula for the lateral area of a pyramid is as follows:

$$L = \frac{1}{2}Pl$$

Remember, P is the perimeter of the base and l is the slant height.

Because the base is a square, the perimeter can be found by multiplying the side length by 4. Substitute into the formula:

$$L = \frac{1}{2}Pl$$

$$L = \frac{1}{2}(4 \cdot 6)(5)$$

$$L = \frac{1}{2}(24)(5)$$

$$L = \frac{1}{2}(120)$$

$$L = 60$$

The lateral area of the pyramid is 60 cm². This can now be used to help find the surface area:

$$S = L + B$$

$$S = \frac{1}{2}Pl + B$$

$$S = \frac{1}{2}Pl + s^2$$

$$S = 60 + 6^2$$

$$S = 60 + 36$$

$$S = 96$$

The surface area of the pyramid in Figure 15.4 is 96 cm².

Finding the surface area of a cone differs from finding the surface area of a pyramid. The formula ($S = L + B$) is the same as for a pyramid, but finding the lateral area and the area of the base is different.

Because the base of every cone is a circle, the area of the base is found by using the area of a circle formula ($A = \pi r^2$). The formula for the lateral area of a cone is as follows (remember, l is the slant height):

$$L = \pi r l$$

Let's try this formula using the cone in Figure 15.5.

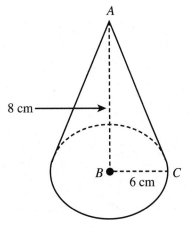

8 cm

6 cm

Figure 15.5: *A cone.*

Like a pyramid, a cone has a right triangle; therefore, you can use the Pythagorean Theorem to find the length of the slant height, $\overline{AC}$:

$$a^2 + b^2 = c^2$$
$$AB^2 + BC^2 = AC^2$$
$$8^2 + 6^2 = AC^2$$
$$64 + 36 = AC^2$$
$$100 = AC^2$$
$$\sqrt{100} = \sqrt{AC^2}$$
$$10 = AC$$

Now that you know the slant height is 10 cm, you can find the lateral area:

$$L = \pi r l$$
$$L = \pi(6)(10)$$
$$L = 60\pi$$

From the lateral area, you can find the surface area:

$$S = L + B$$
$$S = \pi r l + \pi r^2$$
$$S = 60\pi + \pi(6)^2$$
$$S = 60\pi + 36\pi$$
$$S = 96\pi$$
$$S \approx 301.44$$

As you can see, the surface area of the cone is approximately 301.44 cm².

Volume

Volume is the measure of the amount of space inside a figure. Two-dimensional figures do not have an inside; therefore, they do not have volume. Only three-dimensional or solid figures have volume. You find the volume of a figure by determining the number of *unit cubes* that can fit inside it. For this reason, volume is measured in cubic units, or unit³.

> **DEFINITION**
>
> A **unit cube** is a cube with sides that measure 1 unit.

For example, if you are told the volume of a box is 100 in³, it means the box holds 100 cubes with side lengths of 1 inch. If a box has a volume of 50 ft³, it means the box holds 50 cubes with side lengths of 1 foot, and so on.

The Volume of Prisms and Cylinders

The volume of a prism and a cylinder can be found using the following formula:

$$V = Bh$$

B is the area of the base and h is the height.

Where the process differs for prisms and cylinders is in the shape of the base. The base of a prism can have any polygon for a base, such as a rectangle, hexagon, or octagon. Because the bases can differ, the formula for the area of the base can differ.

IT DOESN'T ADD UP

You may have been taught that "volume = length × width × height." While this will work for a rectangular prism or a cube, it is not valid for any other figure.

Let's start with finding the volume of a prism using Figure 15.6.

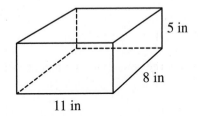

Figure 15.6: *A rectangular prism.*

The solid is a rectangular prism, because the base of the prism is a rectangle. Therefore, when finding the volume, the area of the base, *B*, will be found using the area of the rectangle:

$V = Bh$

$V = (11 \times 8)(5)$

$V = (88)(5)$

$V = 440$

The volume of the rectangular prism is 440 in³.

Remember, the area of the base depends on the shape of the base. If the base of the prism is a triangle, you use the area of a triangle formula $(A = \frac{1}{2}bh)$. If the base of the prism is a polygon with more than four sides, you have to find the apothem and use the formula $[A = \frac{1}{2}aP]$, which you learned in Chapter 11.

Now it's time to learn how to find the volume of a cylinder. Take a look at Figure 15.7.

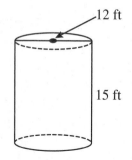

12 ft

15 ft

Figure 15.7: *A cylinder.*

Because all cylinders have a circular base, you can write the volume formula as follows:

$$V = \pi r^2 h$$

Figure 15.7 provides the diameter of the circular base. To use the formula, though, you also need to know the radius. As you know, the radius is half the diameter, so the radius in this case is 6 feet. You can now substitute into the formula:

$$V = Bh$$
$$V = \pi r^2 h$$
$$V = \pi (6)^2 (15)$$
$$V = \pi (36)(15)$$
$$V = 540\pi$$
$$V \approx 1695.6$$

The volume of the cylinder is approximately 1695.6 ft³. You can use this formula for right cylinders, since they also have circular bases.

The Volume of Pyramids and Cones

Imagine you are holding a cone and a cylinder with the same height and base area. If you fill the cone with water and pour it into the cylinder, it will take three full cones of water to fill up the cylinder. The same is true for a prism and a pyramid with the same height and base area. Because of this, the volume of a cone is ⅓ the volume of a cylinder, and the volume of a pyramid is ⅓ the volume of a prism. Therefore, the formula to find the volume of a pyramid or a cone is as follows:

$$V = \frac{1}{3} Bh$$

Just as with prisms and cylinders, the process of finding the area differs depending on the shape of the base.

Pyramids can have any polygon for its base; therefore, the formula for the area of the base depends on the shape. For example, Figure 15.8 shows a triangular pyramid.

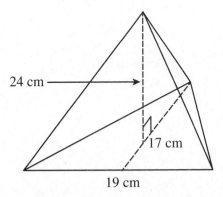

24 cm

17 cm

19 cm

Figure 15.8: *A triangular pyramid.*

Because the base is a triangle, to find the area of the base you need to use the formula for area of a triangle $(A = \frac{1}{2} bh)$. Substitute this information into the volume formula and solve:

$$V = \frac{1}{3} Bh$$

$$V = \frac{1}{3} \left(\frac{1}{2} bh \right) (h)$$

$$V = \frac{1}{3} \left(\frac{1}{2} \cdot 19 \cdot 17 \right) (24)$$

$$V = 1292$$

The volume of the triangular pyramid is 1,292 cm^3.

Let's look at the volume of a cone using Figure 15.9.

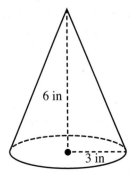

Figure 15.9: *A cone.*

As you learned, cones have circular bases. Because the radius is given, you can substitute directly into the formula:

$$V = \frac{1}{3}Bh$$

$$V = \frac{1}{3}\pi r^2 h$$

$$V = \frac{1}{3}\pi (3)^2 (6)$$

$$V = \frac{1}{3}\pi (9)(6)$$

$$V = 18\pi$$

$$V \approx 56.52$$

The volume of the cylinder is approximately 56.52 in³.

HELPFUL POINT

Because all cones have circular bases, the process for finding the volume is the same. Just be careful to determine whether you are given the radius or the diameter.

Spheres

We've separated out the formulas for spheres from the ones for prisms, cylinders, pyramids, and cones, because unlike those, spheres do not have any bases. The following sections will walk you through the formulas for the surface area and volume of a sphere.

Surface Area

Because they don't have bases, you may think it's difficult to find the surface area of spheres. However, the formula is rather simple:

$$V = 4\pi r^2$$

The only measurement you need to know from a sphere is the radius. In three-dimensional space, the radius of a sphere is the segment connecting the center to any point on a sphere. For example, Figure 15.10 shows you a sphere with its radius indicated.

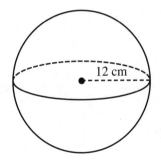

Figure 15.10: *A sphere.*

The sphere in the figure has a radius of 20 cm, so substitute the measure of the radius into the formula to find the surface area:

$$V = 4\pi r^2$$
$$V = 4\pi (12)^2$$
$$V = 4\pi (144)$$
$$V = 576\pi$$
$$V \approx 1808.64$$

The surface area of the sphere is approximately 1,808.64 cm².

Volume

If you recall, volume is the amount of space inside a figure. You can use the following formula to find the volume of a sphere:

$$V = \frac{4}{3}\pi r^3$$

Like the surface area formula, the only measurement you need is the length of the radius. However, be aware that the radius is cubed in this formula.

Let's try this formula using the sphere from Figure 15.10:

$$V = \frac{4}{3}\pi r^3$$

$$V = \frac{4}{3}\pi (12)^3$$

$$V = \frac{4}{3}\pi (1728)$$

$$V = 2304\pi$$

$$V \approx 7234.56$$

The volume of the sphere is approximately 7,234.56 cm^3.

The only way that the process is different from what has been shown here is if the diameter is given. If the diameter is given, you just need to find half of that measure to get the length of the radius before substituting into the formulas.

Similar Solids

Similar solids are solids that have exactly the same shape but not necessarily the same size. Two solids can be proven similar by comparing the ratios of the corresponding measures, known as the *scale factor* (see Chapter 13 for a reminder of how to find scale factor).

Let's take a look at Figure 15.11 and find the scale factor for each pair of similar solids.

The similar solids in 15.11A have a scale factor of $\frac{1}{4}$—the heights are in ratio of $\frac{4}{16}$, which simplifies as $\frac{1}{4}$, while the radii are in ratio of $\frac{2}{8}$, which also simplifies as $\frac{1}{4}$.

The similar solids in 15.11B have a scale factor of $\frac{3}{5}$—the lengths are in ratio of $\frac{3}{5}$; the widths are in ratio of $\frac{9}{15}$, which simplifies as $\frac{3}{5}$; and the heights are in ratio of $\frac{7.5}{12.5}$, which simplifies as $\frac{3}{5}$.

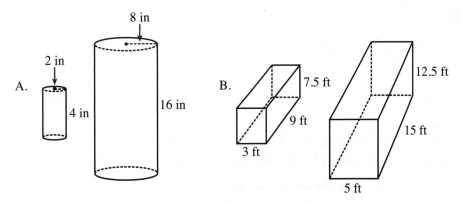

Figure 15.11: *Two pairs of similar solids.*

Sometimes, it will not be clear whether two solids are similar. For this to be true, their scale factor must be in proportion. For example, Figure 15.2 has two cones; how can you tell if they're similar?

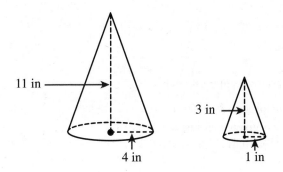

Figure 15.2: *Two cones.*

The ratio of the heights is $\frac{11}{3}$, while the ratio of the radii is $\frac{4}{1}$. Because $\frac{11}{3} \neq \frac{4}{1}$, the solids are not similar.

The Effect of Dilation on Surface Area and Volume

In Chapter 13, you learned that dilation is a reduction or an enlargement of a figure. We are going to show you how dilation can affect the surface area and volume of a solid figure.

In Figure 15.13, Prism A is a reduction of Prism B by a scale factor of $\frac{2}{3}$.

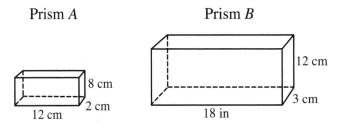

Prism *A*   Prism *B*

8 cm

12 cm

2 cm

3 cm

12 cm   18 in

Figure 15.13: *A pair of similar rectangular prisms.*

You first find the surface area for both using the prism formula you learned earlier in this chapter:

Surface area of Prism $A = Ph + 2B = (28)(8) + 2(24) = 272$ cm^2

Surface area of Prism $B = Ph + 2B = (42)(12) + 2(54) = 612$ cm^2

The ratio of the surface areas is $\frac{272}{612} = \frac{4}{9} = \frac{2^2}{3^2}$. In general, if two solids are similar with a scale factor of $\frac{a}{b}$, then the surface areas have a ratio of $\frac{a^2}{b^2}$. This is shown in Figure 15.13, because the scale factor is $\frac{2}{3}$ and the ratio of the surface areas is $\frac{4}{9}$, or $\frac{2^2}{3^2}$.

Now that you know how to find it for surface area, let's use the rectangular prisms from Figure 15.13 to show you the effect of a dilation on the volume of a solid figure.

Your first step, of course, is finding the volume of both prisms using the formula you learned earlier:

Volume of Prism $A = Bh = (24)(8) = 192$ cm^3

Volume of Prism $B = Bh = (54)(12) = 648$ cm^3

The ratio of the volumes is $\frac{192}{648} = \frac{8}{27} = \frac{2^3}{8^3}$. In general, if two solids are similar with a scale factor of $\frac{a}{b}$, then the volumes have a ratio of $\frac{a^3}{b^3}$. This is shown in Figure 15.13, because the scale factor is $\frac{2}{3}$ and the ratio of the volumes is $\frac{8}{27}$, or $\frac{2^3}{3^3}$.

The Least You Need to Know

- The surface area of a solid figure is the sum of the areas of the figure's faces and curved surfaces.
- Volume is the measure of the space inside a solid figure.
- The surface areas of similar solids are related by the square of the scale factor.
- The volumes of similar solids are related by the cube of the scale factor.

PART 6

Circles

Just like triangles, circles are so complex that they need their own part of this book. Within the language of mathematics, circles provide several new terms and concepts that are necessary for finding measures within them. Whether finding the length of curved or noncurved segments or the measure of a missing angle, this part of the book teaches you all you need to know about the circle.

CHAPTER 16

Properties of Circles

Have you ever tried to define the word *circle?* Some interesting definitions have been given by people learning geometry over the years, such as "360°" or "a figure without sides." We all know what a circle is, but how is it truly defined? In this chapter, we not only help you define what a circle is, but also teach you about all the lines, segments, arcs, and angles associated with circles.

Lines and Segments Related to Circles

A circle is an endless number of points equidistant from a given point. This given point is known as the *center* of the circle. If you were to draw a segment from the center to any point on the circle, this segment would be called the *radius*. Any given circle has an infinite number of radii.

Let's look at some more segments and lines related to circles in Figure 16.1.

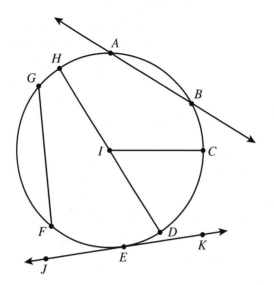

Figure 16.1: *Can you pinpoint the radius, diameter, chord, secant, and tangent?*

First, let's identify the center and the radius. The center of the preceding circle is located at point *I.* The segment that connects point *I* and point *C* is the radius, which is called $\overline{IC}$ in this figure.

$\overline{HD}$ is the *diameter*; the length of the diameter is twice the length of the radius. A diameter is a special type of chord or segment that connects two points on a circle. Another example of a chord on the circle is $\overline{GF}$.

 **DEFINITION**

A **diameter** is a segment that connects two points on the circle and passes through the center. A diameter is the longest chord in a circle.

The circle in Figure 16.1 also has two lines, $\overleftrightarrow{AB}$ and $\overleftrightarrow{JK}$. $\overleftrightarrow{AB}$ is a secant, which is a line that intersects a circle at two points. Secants are both inside and outside of a circle. $\overleftrightarrow{JK}$ is a tangent, which is a line that intersects a circle at exactly one point. That point of intersection is called the *point of tangency.*

Properties of Tangents

You already know that a tangent touches a circle in exactly one place, but there are two other properties also associated with it. The first property is that a tangent is perpendicular to the radius of the circle. Because the tangent and the radius are perpendicular, their intersection forms a 90-degree angle. If the tangent is drawn as a segment, this property can be used to find the length of the tangent or the radius. Take a look at Figure 16.2 and then try to find the length of the segment.

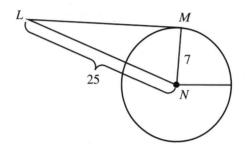

Figure 16.2: $\overline{LM}$ *is a tangent to the circle with center N.*

To be a tangent, $\overline{LM}$ must intersect $\overline{MN}$ at 90°. Therefore, ΔLMN must be a right triangle. Because ΔLMN is a right triangle, the Pythagorean Theorem can be used to find the length of the tangent. So substitute the values in Figure 16.2 into the Pythagorean Theorem:

$$a^2 + b^2 = c^2$$
$$(LM)^2 + 7^2 = 25^2$$
$$(LM)^2 + 49 = 625$$
$$(LM)^2 + 49 - 49 = 625 - 49$$
$$(LM)^2 = 576$$
$$\sqrt{(LM)^2} = \sqrt{576}$$
$$LM = 24$$

The length of the tangent is 24.

Another property of tangents is that if two tangents are drawn from a common point outside a circle, they are congruent. For example, in Figure 16.3, two tangents are drawn from point *A*.

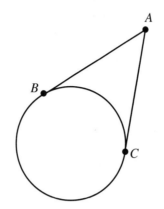

Figure 16.3: *Two tangents drawn from a common point outside the circle.*

$\overline{AB}$ is a tangent whose point of tangency is point *B*, while $\overline{AC}$ is a tangent whose point of tangency is point *C*. Because both tangents begin at point *A*, $\overline{AB} \cong \overline{AC}$.

Often you will see this second property used in situations with algebraic equations. For example, using Figure 16.3, assume that $AB = 25$ and $AC = 2x + 3$. Knowing that $\overline{AB} \cong \overline{AC}$, the following equation can be used to solve for *x*:

$$2x + 3 = 25$$
$$2x + 3 - 3 = 25 - 3$$
$$2x = 22$$
$$\frac{2x}{2} = \frac{22}{\cdot 2}$$
$$x = 11$$

So $x = 11$. As you can see, the congruency of the two segments allows equations to be set up and solved.

Arcs and Angles

In circles, certain relationships exist between arcs and angles. Arcs, which are segments of a circle, are measured based on the central angle or inscribed angle that intercepts the arc. First, though, let's learn about the different types of arcs and angles associated with circles. We'll then take you through equations involving arcs and angles.

Central Angles and Inscribed Angles

A central angle is an angle whose vertex is located at the center of a circle, while an inscribed angle is an angle whose vertex is on a circle and whose sides each intersect the circle at another point. Figure 16.4 shows both types of angles.

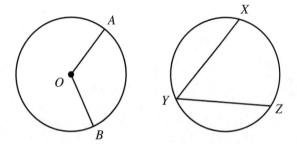

Figure 16.4: *The left circle shows a central angle, while the right circle shows an inscribed angle.*

In the circle on the left, $\angle AOB$ is a central angle; the vertex of the angle is located at the center, O. In the circle on the right, $\angle XYZ$ is an inscribed angle; the vertex, Y, is a point on the circle.

> **HELPFUL POINT**
>
> A central angle is formed by two radii, while an inscribed angle is formed by two chords.

Using Figure 16.5, let's look at an example of how to find measure of a central angle.

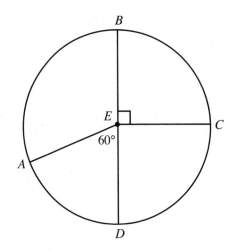

Figure 16.5: *Multiple central angles are drawn with vertices at center E.*

In Figure 16.5, $\overline{BD}$ is a diameter; it forms a straight angle, $\angle BED$, and measures 180°. $\angle BEC$ is a right angle and measures 90°.

To find the measure of $\angle AEB$, you need to realize that $\angle AEB$ and $\angle AED$ are a linear pair, because together they form the straight angle $\angle BED$. Because they are a linear pair, the angles are supplementary. Therefore, you can use the following to find $\angle AEB$:

$$m\angle AEB + m\angle AED = m\angle BED$$

$$m\angle AEB + 60° = 180°$$

$$m\angle AEB + 60° - 60° = 180° - 60°$$

$$m\angle AEB = 120°$$

The central angle measures 120°.

Types of Arcs

Think about the crust of a slice of pizza. This crust would be considered an arc, because the crust on a single slice is a part of the entire circular crust of the pizza.

Figure 16.6 shows you different examples of arcs.

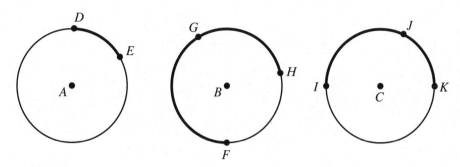

Figure 16.6: *Each circle displays an arc of varying measure.*

In circle *A*, $\overset{\frown}{DE}$ is a minor arc. Minor arcs measure less than 180° and are named by the endpoints of the arc. In circle *B*, $\overset{\frown}{FGH}$ is a major arc. Major arcs measure more than 180° and are named by the endpoints and another point on the arc. In circle *C*, $\overset{\frown}{IJK}$ measures exactly 180°, which makes it a semicircle. Semicircles are also named by the endpoints and another point on the arc.

Determining the Measure of an Arc with Angles

The exact measure of an arc can be determined by the measure of its central angle or inscribed angle. The measure of the central angle is equal to the measure of its *intercepted arc*, while the measure of the inscribed angle is equal to half of the measure of the intercepted arc.

 **DEFINITION**

An **intercepted arc** is the part of the circle between the intersection points.

First, let's look more closely at the relationship between a central angle and its intercepted arc using Figure 16.7. For the circle, find the $m\overarc{AD}$ and $m\overarc{AB}$.

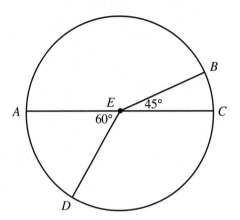

Figure 16.7: *Central angles are drawn with vertices at center E.*

$\overarc{AD}$ is the intercepted arc of the central angle $\angle AED$. Because the central angle measures 60°, the intercepted arc has the same measure: $m\overarc{AD} = 60°$.

$\overarc{AB}$ is the intercepted arc of the central angle $\angle AEB$. $\angle AEB$ and $\angle BEC$ are supplementary angles, because they form the straight angle $\angle AEC$. So you can use the following to find $m\overarc{AB}$:

$$m\angle AEB + m\angle BEC = m\angle AEC$$
$$m\angle AEB + 45° = 180°$$
$$m\angle AEB + 45° - 45° = 180° - 45°$$
$$m\angle AEB = 135°$$

Because $m\angle AEB = 135°$ and it intercepts $\overarc{AB}$, $m\overarc{AB} = 135°$.

Now it's time to examine the relationship between inscribed angles and their intercepted arcs. Unlike central angles, an inscribed angle is equal to half of the measure of its intercepted arc. Let's try to apply this using the information in Figure 16.8.

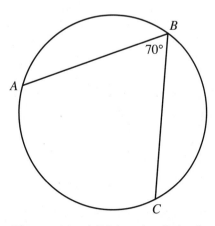

Figure 16.8: $\angle ABC$ *is an inscribed angle.*

Because $\angle ABC$ is an inscribed angle, $m\angle ABC = \frac{1}{2}m\widehat{AC}$. Here's the equation you use:

$$m\angle ABC = \frac{1}{2}m\widehat{AC}$$

$$70° = \frac{1}{2}m\widehat{AC}$$

$$2(70°) = 2\left(\frac{1}{2}m\widehat{AC}\right)$$

$$140° = m\widehat{AC}$$

Because $m\angle ABC = 70°$ and it intercepts $\widehat{AC}$, $\widehat{AC} = 140°$.

But what if you don't know the measure of the inscribed angle? For example, say you have to find $m\angle EGF$ for Figure 16.9.

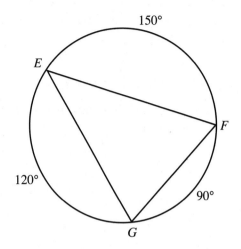

Figure 16.9: *Three inscribed angles are drawn in the circle.*

HELPFUL POINT

In Figure 16.9, the inscribed angles form a triangle. This is an example of an inscribed polygon, which is a polygon inside a circle with all the vertices on the circle.

To find $m\angle EGF$, you need to know which arc it intercepts. The segments of the angle intersect points E and F; therefore, $m\angle EGF = \frac{1}{2}m\widehat{EF}$. Substitute into the equation:

$$m\angle EGF = \frac{1}{2}m\widehat{EF}$$

$$m\angle EGF = \frac{1}{2}\left(150°\right)$$

$$m\angle EGF = 75°$$

Because $\angle EGF$ intersects $\widehat{EF}$, $m\angle EGF = 75°$.

Theorems Associated with Chords

Chords are segments that occur inside a circle. When two of these chords intersect, they form relationships between their lengths and the angles created. These relationships result in two theorems:

Lengths of Intersecting Chords Theorem

When two chords intersect, the point of intersection splits each chord into two smaller segments. When the lengths of these segments are multiplied, the products are equal. Let's take a look at how this theorem works in the circle in Figure 16.10.

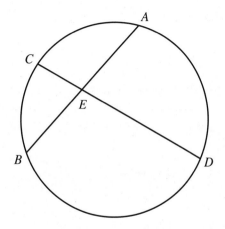

Figure 16.10: *Two intersecting chords drawn inside a circle.*

Based on the Lengths of Intersecting Chords Theorem, the product of *AE* and *EB* is equal to the product of *CE* and *ED* shown by the following equation:

$$AE \cdot EB = CE \cdot ED$$

You can now use this equation to find unknown segment lengths. Referring to Figure 16.10 again, let's assume that $AE = 6$ inches, $EB = 5$ inches, and $ED = 10$ inches. Substitute into the equation to find the length of CE:

$$AE \cdot EB = CE \cdot ED$$

$$5 \cdot 6 = CE \cdot 10$$

$$30 = CE \cdot 10$$

$$\frac{30}{10} = \frac{CE \cdot 10}{10}$$

$$3 = CE$$

The length of CE is 3 inches.

Angles Formed by Intersecting Chords Theorem

This theorem states that if two chords intersect inside a circle, the measure of each angle is one half the sum of the arcs intercepted by the angle and its vertical angle. That may be hard to understand as written out, so let's look at the relationship in Figure 16.11.

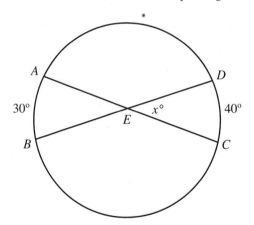

Figure 16.11: *Two intersecting chords drawn in a circle.*

In Figure 16.11, you have to find the value of x, which is the measure of $\angle DEC$. $\angle DEC$ intercepts $\overset{\frown}{DC}$ and its vertical angle, $\angle AEB$, intercepts $\overset{\frown}{AB}$. Based on the Angles Formed by Intersecting Lines Theorem, the following equation can be used to solve for x:

$$x = \frac{1}{2}\left(m\overset{\frown}{AB} + m\overset{\frown}{DC} \right)$$

Substitute the values from Figure 16.11 into the equation to find *x:*

$$x = \frac{1}{2}\left(m\widehat{AB} + m\widehat{DC} \right)$$

$$x = \frac{1}{2}\left(30° + 40° \right)$$

$$x = \frac{1}{2}\left(70° \right)$$

$$x = 35°$$

Remember, vertical angles are congruent; therefore, $m\angle AEB = 35°$.

> **HELPFUL POINT**
>
> If only one vertical angle pair is known, you need to use supplementary angles. $\angle DEC$ is supplementary to $\angle AED$ and $\angle CEB$ because they are linear pairs; therefore, $m\angle AED = m\angle CEB = 145°$.

Angles Formed by Secants and Tangents

When two tangents, two secants, or one secant and one tangent intersect, they form an angle outside the circle. The measure of this angle is equal to half of the difference of the intercepted arcs. You can see an example of each type in Figure 16.12.

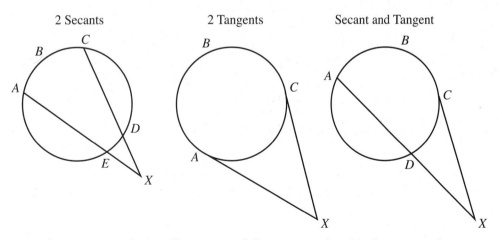

Figure 16.12: *Angles formed by two secants (left), two tangents (center), and one secant and one tangent (right).*

The secants and tangents in each of the circles intersect to form an angle outside the circle—$\angle X$. Each line intersects the circle; these intersections form arcs. The intercepted arcs can be thought of as a far arc and a near arc. In any of the three circles, to find $m\angle X$, it is half the difference of the far arc and the near arc:

$$m\angle X = \frac{1}{2}\left(\textit{far arc} - \textit{near arc}\right)$$

HELPFUL POINT

By a far arc, we mean the intercepted arc farthest from the angle. The near arc is the one closest to the angle.

For example, in the left circle, which is intersected by two secants, $\overset{\frown}{AC}$ is the arc far from $\angle X$ and $\overset{\frown}{DE}$ is the arc near $\angle X$. Therefore, the equation would be as follows:

$$m\angle X = \frac{1}{2}\left(m\overset{\frown}{AC} - m\overset{\frown}{DE}\right)$$

Let's assume that $m\overset{\frown}{AC} = 80°$ and $m\overset{\frown}{DE} = 30°$. Substitute these values into the equation:

$$m\angle X = \frac{1}{2}\left(m\overset{\frown}{AC} - m\overset{\frown}{DE}\right)$$

$$m\angle X = \frac{1}{2}\left(80° - 30°\right)$$

$$m\angle X = \frac{1}{2}\left(50°\right)$$

$$m\angle X = 25°$$

Therefore, $m\angle X = 25°$.

The same process can be used whether the angle is formed by two secants, two tangents, or one secant and one tangent.

The Least You Need to Know

- A circle is an endless number of points equidistant from a given point.
- A tangent touches a circle in exactly one place and is perpendicular to the radius of a circle. If two tangents are drawn from a common point outside a circle, they are congruent.
- Chords are segments that occur inside a circle. When two chords intersect, the relationships they form can be explained by two theorems: the Lengths of Intersecting Chords Theorem and the Angles Formed by Intersecting Chords Theorem.
- The measure of an angle formed by the intersection of two tangents, two secants, or one secant and one tangent is equal to half of the difference of the intercepted arcs.

CHAPTER 17

Circle Circumference and Area

Circles are two-dimensional figures; however, they do not have any sides, meaning they do not have a perimeter. Still, a circle does have measurements for the length around and the surface of a circle, which are known as *circumference* and *area*. In this chapter, we show you how to find the circumference and area of a circle, as well as parts of these measures.

Circumference of a Circle

Circumference is the distance around the outside of a circle measured in units. If you took a string and wrapped it around a circle, the length of the string needed is the circumference of the circle. Because it's silly to carry string and a ruler around to measure circles, there are two formulas that can be used to find circumference:

$$C = \pi d \text{ or } C = 2\pi r$$

In the formulas, C represents the circumference, d represents the length of the diameter, and r represents the length of the radius. The two formulas are equivalent because the diameter is equal to twice the radius. Which formula you use depends on the information given about the circle.

In This Chapter

- Finding the circumference and area of a circle
- Measuring sectors and solving for the length of arcs
- Which formula to apply when solving a circle measurement problem

Let's take a look at the two circles in Figure 17.1.

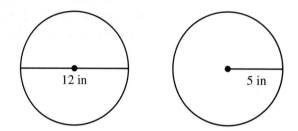

Figure 17.1: *The circle on the left gives the length of the diameter, while the circle on the right gives the length of the radius.*

In Figure 17.1, the left circle has a diameter that measures 12 inches. For this circle, you use the formula $C = \pi d$, because you know the length of the diameter. Substitute the length of the diameter into the formula:

$$C = \pi d$$
$$C = \pi(12)$$
$$C = 12\pi$$
$$C \approx 37.68$$

The circumference of the first circle is approximately 37.68 inches.

HELPFUL POINT

Throughout this chapter, we multiply by 3.14 to get an approximate number for the answer. However, leaving an answer in terms of π is more precise.

Going back to Figure 17.1, the second circle has a radius that measures 5 inches. For this circle, I will use $C = 2\pi r$ to find the circumference. Substitute into the formula.

$$C = 2\pi r$$
$$C = 2\pi(5)$$
$$C = 10\pi$$
$$C \approx 31.4$$

The circumference of the second circle is approximately 31.4 inches.

The same circumference formulas can be used to find the length of a diameter or a radius when the circumference is given. For example, if the circumference of a circle is 8π feet or approximately 25.26 feet, you can use a circumference formula to find the radius and diameter of the circle. In this case, we're going to use $C = \pi d$, but you can use either formula to solve:

$$C = \pi d$$
$$8\pi = \pi d$$
$$\frac{8\pi}{\pi} = \frac{\pi d}{\pi}$$
$$8 = d$$

The diameter of the circle is 8 feet. If you recall, radius is half the length of diameter, so the radius for this circle is 4 feet.

Arc Length

In Chapter 16, you learned about arcs, which are segments of a circle. In order to find the length of an arc, think of it as being a piece of the circumference

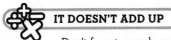 **IT DOESN'T ADD UP**

Don't forget—arcs have both a degree measure *and* a length.

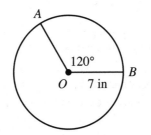

Figure 17.2: $\angle AOB$ *is drawn in circle O.*

In Figure 17.2, $\angle AOB$ is a central angle, because the vertex is on the center of the circle. From Chapter 16, you know the measure of an arc is equal to the measure of its central angle; therefore, $m\overarc{AB}$ is 120°. Now that you have this information, we're going to show you how to find the length of $\overarc{AB}$.

First, you need to find the circumference. Looking back at the circle in Figure 17.2, the circumference of the circle can be found by substituting the radius into one of the circumference formulas; here, we use $C = 2\pi r$:

$$C = 2\pi r$$
$$C = 2\pi(7)$$
$$C = 14\pi$$
$$C \approx 43.96$$

If the circumference of the circle is approximately 43.96 inches, the length of $\overparen{AB}$ must be a fraction of this length. You must know the exact fraction of the circumference that $\overparen{AB}$ represents. Because an entire circle is 360° and $m\overparen{AB} = 120°$, $\overparen{AB}$ is $\dfrac{120°}{360°}$ or $\dfrac{1}{3}$ of the circumference. Therefore, the length of $\overparen{AB} = \dfrac{1}{3}(43.96) \approx 14.7$.

Based on this reasoning, you can create a formula for finding the length of an arc. For the circle in Figure 17.2, the formula is the following:

$$\overparen{AB} = \frac{m\angle AOB}{360°} \cdot 2\pi r$$

Let's look at one more example in Figure 17.3 to see how the formula is used. Using the information, find the length of $\overparen{CD}$.

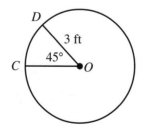

Figure 17.3: $\angle COD$ *is drawn in circle O.*

To find the length of the arc, substitute into the formula:

$$\frac{45°}{360°} \cdot 2\pi(3)$$

$$\frac{1}{8} \cdot 6\pi$$

$$\frac{3}{4}\pi$$

$$\approx 2.4$$

The length of $\overset{\frown}{CD}$ is approximately 2.4 feet.

Area of a Circle

The area of a circle is the measurement of the surface of the circle measured in square units. The formula to find the area of a circle is the following:

$$A = \pi r^2$$

In the formula, A represents area and r represents the length of the radius. Try using this formula to find the area of the circle in Figure 17.4.

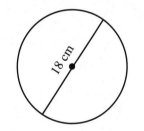

Figure 17.4: *A diameter is drawn in the circle.*

In Figure 17.4, the length of the diameter is 18 centimeters. Because the area formula uses the length of the radius (which is half the diameter), take half of the diameter. This gives you a radius of 9 centimeters. Now substitute into the formula:

$$A = \pi r^2$$
$$A = \pi (9)^2$$
$$A = 81\pi$$
$$A \approx 254.34$$

The area of the circle is approximately 254.34 square centimeters.

The area formula for a circle can also be used to find the length of a radius. When the area is given, you can work backward to find the radius. For example, if the area of a circle is approximately 50.24 square feet, you can substitute into the area formula and solve for the radius:

$$A = \pi r^2$$
$$50.24 = \pi r^2$$
$$\frac{50.24}{\pi} = \frac{\pi r^2}{\pi}$$
$$16 = r^2$$
$$\sqrt{16} = \sqrt{r^2}$$
$$4 = r$$

IT DOESN'T ADD UP

When solving $16 = r^2$, remember to find the square root of both sides. Often people have a tendency to divide both sides by 2.

The radius of the circle is 4 feet long.

Area of a Sector

Remember when we told you in Chapter 16 that an arc is like the crust on a slice of pizza? Well, the slice of pizza itself can be thought of as a representation of a *sector*. A sector is a fraction of a whole circle. In Figure 17.5, the shaded section of the circle is a sector.

📖 DEFINITION

A **sector** is a portion of a circle enclosed by two radii and the intercepted arc.

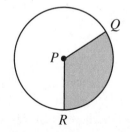

Figure 17.5: *Circle P has a portion of the circle shaded. The shaded part is a sector.*

In the figure, the shaded region is called sector *QPR*. It is enclosed by the radii $\overline{PQ}$ and $\overline{PR}$, as well as $\overset{\frown}{QR}$. Because it is less than half of the circle, sector *QPR* is known as a *minor sector;* the unshaded region is considered a *major sector.*

There are two special types of sectors:

- A sector that is a quarter of a circle is called a *quadrant.*

- A sector that is half of a circle is called a *semicircle.*

The area of a sector is a fraction of the area of the entire circle. To find the area of a sector, you first need to find the area of the entire circle. Let's look at Figure 17.6.

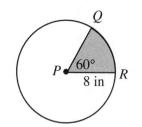

Figure 17.6: *For circle P, the shaded sector is a minor sector.*

The radius of circle *P* in the figure is 8 inches. Substitute the radius into the area formula to first find the area of the circle:

$$A = \pi r^2$$

$$A = \pi(8)^2$$

$$A = 64\pi$$

$$A \approx 200.96$$

The area of the circle is approximately 200.96 square inches. The area of the shaded sector is a fraction of that area; however, you must know the exact fraction of the area that minor sector *QPR* represents. Because an entire circle is 360° and sector *QPR* is 60°, sector *QPR* is $\frac{60°}{360°}$ or $\frac{1}{6}$ of the circle. Therefore, the area of sector $QPR = \frac{1}{6}(200.96) \approx 33.5$.

Based on this reasoning, you can create a formula for finding the area of a sector. For the circle in Figure 17.6, the formula is the following:

$$\frac{m\angle QPR}{360°} \cdot \pi r^2$$

Let's look at one more example, Figure 17.7, to see how the formula is used. Find the area of the shaded sector *LMN*.

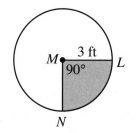

Figure 17.7: *For circle M, the shaded sector is a quadrant.*

To find the area of sector *LMN*, substitute into the formula:

$$\frac{90°}{360°} \cdot \pi(3)^2$$

$$\frac{1}{4} \cdot 9\pi$$

$$\frac{9}{4}\pi$$

$$\approx 7.1$$

The area of sector *LMN* is approximately 7.1 square feet.

Application of Circumference and Area

When working with circles, we have shown you how to find circumference and area. While the formulas make it easy to find these measurements, it is often difficult to determine which measurement you need in a problem situation.

> **HELPFUL POINT**
>
> When solving word problems, remember that circumference is the length around the outside of a figure, while area is the space that covers the figure.

For example, Jake has a pool in his backyard. For safety purposes, he needs to buy a fence to enclose his pool. He also wants to replace the tiles on the floor of his pool. Before he can go shopping, he needs to know the amount of fencing and tiles needed for his pool, as shown in Figure 17.8. Find how much Jake needs.

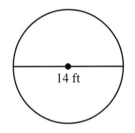

Figure 17.8: *Jake's pool.*

To find the correct amount of fencing and tiles, you need to determine whether you have to find the circumference or the area. Earlier in this chapter, we said that circumference is the length around a circle; this can help you determine the amount of fencing needed, because it will be going around the pool. We also said that area is the measure of the surface of the circle; to know how many tiles to buy, you have to know the measure of the surface of the pool bottom.

First, to find the amount of fencing, substitute the diameter into one of the circumference formulas:

$$C = \pi d$$
$$C = \pi(14)$$
$$C \approx 43.96$$

Jake will need approximately 43.96 feet of fencing to surround his pool.

For the tiles, you need to know the area of the bottom of the pool. The radius of the pool is half of 14, which is 7. Substitute the radius into the area formula:

$$A = \pi r^2$$
$$A = \pi (7)^2$$
$$A = 49\pi$$
$$A \approx 153.86$$

The tiles need to cover approximately 153.86 square feet.

When working with circumference and area, remember that circumference surrounds a figure and area covers a figure. Keeping these facts in mind will help you determine the measurement you need to use to solve a problem involving measurements of a circle.

The Least You Need to Know

- Two formulas can help you find the circumference of a circle: $C = \pi d$ and $C = 2\pi r$.
- The length of an arc is a fraction of the circumference of a circle.
- The formula to find the area of a circle is $A = \pi r^2$.
- The area of a sector is a fraction of the total area of a circle.
- Circumference measures the surrounding length of a circle, while area measures the space that covers a circle.

Coordinate Geometry

What would happen if you sliced a cone? Depending on how you sliced it, several different shapes can be formed. These resulting shapes can be graphed on the coordinate plane and have equations that represent them. This part explores these shapes and takes the basics that you have already learned in Part 1 of this book to expand the way you think about the coordinate plane. Along with the shapes formed by slicing a cone, you study the movement of figures on the coordinate plane.

CHAPTER 18

A Deeper Look at Coordinate Geometry

Did you ever see a cone and think about slicing it? Probably not, but in mathematics, it is common to slice a cone. Mathematicians refer to this as a cross-section of a right circular cone. When a cone is intersected by a plane, the resulting cross-sections are circles, parabolas, ellipses, or hyperbolas. In this chapter, we show you how these cross-sections occur, graph them on the coordinate plane, and find their equations.

Conic Sections

A cross-section of a right circular cone due to the intersection of a cone with a plane is called a *conic section*. Depending on the angle of the plane intersecting the cone, different shapes are formed, such as circles, parabolas, and ellipses.

In This Chapter

- Exploring conic sections
- Graphing circles, parabolas, and ellipses on a coordinate plane
- Identifying the parts associated with the graph of each conic section
- Finding the equations of circles, parabolas, and ellipses

On a coordinate plane, a circle is defined by its center and radius. An ellipse is an elongated circle with two perpendicular axes of different lengths. When an ellipse is centered around the origin, the x-axis and y-axis are the perpendicular axes. A parabola is the set of all points equidistant from a given point on the coordinate plane; it is a single curve. The equations of most conic sections contain both x^2 and y^2, except a parabola, which contains only one of the two.

Circles

A circle is one type of cross-section of a right circular cone. It is formed by the intersection of a cone with a plane that is perpendicular to the height of the cone. For example, the shaded cross-section in Figure 18.1 is a circle.

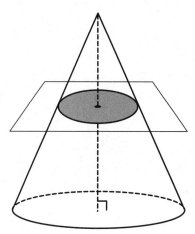

Figure 18.1: *A circular cross-section.*

Writing the Equation for a Circle

The standard form for the equation of a circle, which is used to define a circle on the coordinate plane, is as follows:

$$(x - h)^2 + (y - k)^2 = r^2$$

In the equation, h is the x-coordinate of the center, k is the y-coordinate of the center, and r is the radius of the circle.

First, let's look at how to write an equation for a circle given the center and the radius. If the center of a circle is located at (-3, 0) and the radius is 7 units, you can substitute these values into the standard form for the equation of a circle:

$$(x - h)^2 + (y - k)^2 = r^2$$

$$(x - [-3])^2 + (y - 0)^2 = 7^2$$

$$(x + 3)^2 + y^2 = 49$$

Because the center is located at (-3, 0), $h = -3$ and $k = 0$. With a radius of 7, the equation of the circle is $(x + 3)^2 + y^2 = 49$.

Now try writing an equation using the circle graphed in Figure 18.2.

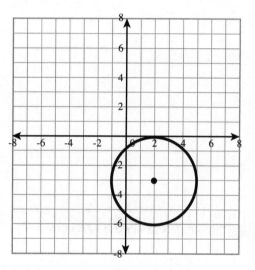

Figure 18.2: *A circle on the coordinate plane.*

Before you can write the equation, you need to identify the center of the circle and the length of the radius. The center is located at (2, -3). Therefore, $h = 2$ and $k = -3$. The radius can be found on the graph by starting at the center and counting the space to an outer point on the circle. If you do that, you'll find the radius is 3 units.

HELPFUL POINT

Remember, when counting spaces on a coordinate plane, you can only count horizontally or vertically. Counting on a diagonal will not give you an accurate answer.

Now it's time to substitute the values from the graph into the equation:

$$(x - h)^2 + (y - k)^2 = r^2$$

$$(x - 2)^2 + (y - [\text{-}3])^2 = 3^2$$

$$(x - 2)^2 + (y + 3)^2 = 9$$

The equation of the circle graphed on the coordinate plane is $(x - 2)^2 + (y + 3)^2 = 9$.

Using Other Information to Write a Circle Equation

For each circle you have looked at so far, you have been given the center and the radius. However, it is possible to write the equation of a circle given different pieces of information.

Let's say you know that the center of a circle is located at (-2, 4) and the circle passes through the point (-6, 7). So you have the center of the circle, but you do not have the length of the radius. Remember Chapter 4, when we spoke about the distance formula? You can use it to find the distance from the center to the point on the circle, which will give you the radius.

Using (-2, 4) as (x_1, y_1) and (-6, 7) as (x_2, y_2), substitute into the distance formula:

$$d = \sqrt{\left(x_2 - x_1\right)^2 + \left(y_2 - y_1\right)^2}$$

$$d = \sqrt{\left(-6 - (-2)\right)^2 + \left(7 - 4\right)^2}$$

$$d = \sqrt{\left(-4\right)^2 + \left(3\right)^2}$$

$$d = \sqrt{16 + 9}$$

$$d = \sqrt{25}$$

$$d = 5$$

The distance from the center to a point on the circle is 5 units; therefore, the length of the radius is 5 units. With this information, you have what you need to write the equation of the circle:

$$(x - h)^2 + (y - k)^2 = r^2$$

$$(x - [\text{-}2])^2 + (y - 4)^2 = 5^2$$

$$(x + 2)^2 + (y - 4)^2 = 25$$

The equation of the circle with a center at (-2, 4) and that passes through (-6, 7) is $(x + 2)^2 + (y - 4)^2 = 25$.

Graphing a Circle

Now it's time to graph a circle given the equation. To be able to graph, you need to know the center of the circle and the length of the radius. In the standard form of the equation, you can obtain this information directly from the equation. Take a look at the following:

$$x^2 + (y + 1)^2 = 36$$

The center of the circle is located at the point (h, k) given the equation $(x - h)^2 + (y - k)^2 = r^2$. Therefore, $x^2 + (y + 1)^2 = 36$ results in $(x - 0)^2 + (y - [-1])^2 = 6^2$. Once back in its original form, you can see that the center of the circle (h, k) is $(0, -1)$ and the radius is 6 (the square root of 36). You can now use this information to graph the circle, which you can see in Figure 18.3.

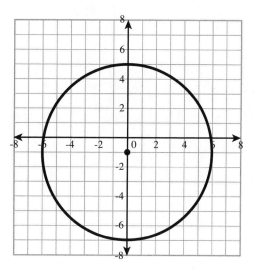

Figure 18.3: *A circle graphed on the coordinate plane.*

Parabolas

A parabola is another type of conic section. It is the cross-section formed by the intersection of a cone and a plane that is parallel to the edge of a cone. In Figure 18.4, the shaded section is the result of this intersection.

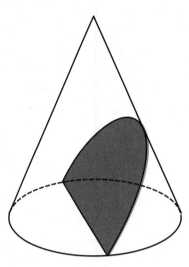

Figure 18.4: *A parabolic cross-section.*

When a parabola is graphed on a coordinate plane, it is the set of all points that are an equal distance from both the *focus* and the *directrix,* as you can see in Figure 18.5. The parabola in the figure has a vertex at (0, 0); a focus, *F,* located at (-6, 0); and a directrix that's the line $x = 6$.

> **DEFINITION**
>
> A **focus** is a fixed point, *F*, used to define a parabola. A **directrix** is a fixed line that is also used to define a parabola.

To determine the equation of the parabola in Figure 18.5, we have to give you some information. There are two forms of the standard form for the equation of a parabola with a vertex at (0, 0). One equation describes a parabola that opens right or left, while the other equation describes a parabola that opens up or down.

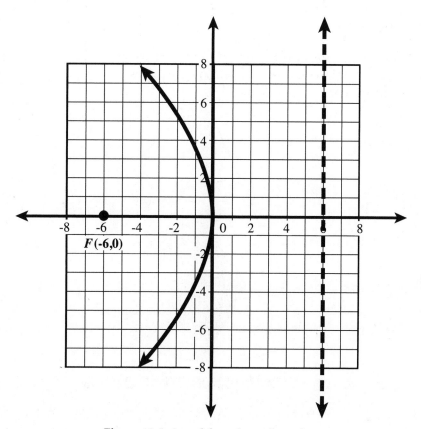

Figure 18.5: *A parabola on the coordinate plane.*

Writing the Equation for a Parabola That Opens Right or Left

The standard form for the equation of a parabola that opens right or left and has a vertex at (0, 0) is the following:

$$x = \frac{1}{4p} y^2$$

In this equation, the parabola opens to the right if p is positive and opens left if p is negative. The focus is located at $(p, 0)$, and the directrix is the line $x = -p$.

Looking at the parabola in Figure 18.5, the parabola opens to the left. As you know, the focus is given at (-6, 0) and the directrix is at $x = 6$. Because the focus is at (-6, 0), the value of p is -6. Substitute $p = -6$ into the equation:

$$x = \frac{1}{4p} y^2$$

$$x = \frac{1}{4(-6)} y^2$$

$$x = -\frac{1}{24} y^2$$

The equation of the parabola in Figure 18.5 is $x = -\frac{1}{24} y^2$.

IT DOESN'T ADD UP

The negative sign doesn't just apply to the denominator. Therefore, instead of keeping the negative sign there, make sure it moves to the front of the fraction.

Writing the Equation for a Parabola That Opens Up or Down

The standard form for the equation of a parabola that opens up or down and has a vertex at (0, 0) is as follows:

$$y = \frac{1}{4p} x^2$$

In this equation, the parabola opens right if p is positive and opens down if p is negative. The focus is located at (0, p) and the directrix is the line $y = -p$.

Let's say a parabola has a focus at (0, 5) and a directrix at y = -5. Because the focus is on the y-axis and the directrix is a horizontal line, the parabola will open up or down. The directrix occurs below the focus on the coordinate plane, so the parabola must open up. Because the focus is (0, 5), p = 5. Substitute p = 5 into the standard form of the equation:

$$y = \frac{1}{4p}x^2$$

$$y = \frac{1}{4(5)}x^2$$

$$y = \frac{1}{20}x^2$$

The equation of the parabola is $y = \frac{1}{20}x^2$.

HELPFUL POINT

In algebra, a parabola is the graph of a quadratic function. However, those parabolas only open up or down. A parabola that opens left or right can't be a function because each x-value has more than one y-value.

Ellipses

An ellipse is a conic section formed by the intersection of a cone and a slightly angled plane. In Figure 18.6, the shaded section is the result of this intersection.

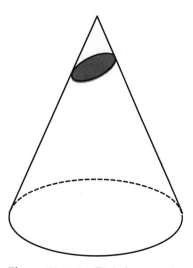

Figure 18.6: *An elliptical cross-section.*

On the coordinate plane, an ellipse has two axes. The longer axis of an ellipse is called the *major axis*, and the shorter axis of an ellipse is called the *minor axis*. The endpoints of the major axis are called the *vertices*, and the endpoints of the minor axis are called the *co-vertices*.

The standard form of an ellipse centered at (0, 0) depends on whether the major axis is horizontal or vertical. If the major axis is horizontal, the ellipse appears that it is being stretched left to right. If the major axis is vertical, it appears that it is being stretched up and down.

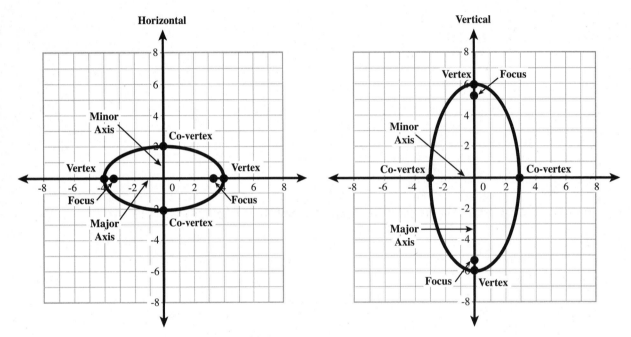

Figure 18.7: *Two ellipses—one with a horizontal major axis and one with a vertical major axis.*

Writing the Equation for an Ellipse with a Horizontal Axis

The standard form of the equation for an ellipse with a horizontal major axis is the following:

$$\frac{x^2}{a^2} + \frac{y^2}{b^2} = 1$$

The *a* is the *x*-coordinates of the vertices and the *b* is the *y*-coordinate of the co-vertices. In the top graph of Figure 18.7, the *x*-coordinates of the vertices are -4 and 4. Because the equation squares the value of *a*, either value can be used. The *y*-coordinates of the co-vertices are -2 and 2.

Again, the equation squares the value of b, so either value can be used. Substitute the values of a and b into the equation:

$$\frac{x^2}{a^2} + \frac{y^2}{b^2} = 1$$

$$\frac{x^2}{4^2} + \frac{y^2}{2^2} = 1$$

$$\frac{x^2}{16} + \frac{y^2}{4} = 1$$

The equation of the ellipse is $\dfrac{x^2}{16} + \dfrac{y^2}{4} = 1$.

 **HELPFUL POINT**

Remember, the square of a negative number is always positive.

Writing the Equation for an Ellipse with a Vertical Axis

Now let's look at the second graph from Figure 18.7, which has a vertical major axis. The standard form of the equation for an ellipse with a vertical major axis is as follows:

$$\frac{y^2}{a^2} + \frac{x^2}{b^2} = 1$$

For this equation, a is the y-coordinates of the vertices and b is the x-coordinates of the co-vertices. In the graph, a is -6 or 6 and b is -3 or 3. Remember, you can use either value. Substitute the values of a and b into the equation:

$$\frac{y^2}{a^2} + \frac{x^2}{b^2} = 1$$

$$\frac{y^2}{6^2} + \frac{x^2}{3^2} = 1$$

$$\frac{y^2}{36} + \frac{x^2}{9} = 1$$

The equation of the ellipse is $\dfrac{y^2}{36} + \dfrac{x^2}{9} = 1$.

Using Foci to Find Missing Values

On the graphs from Figure 18.7, you can also see the label "focus"; these points are known collectively as *foci*. The foci are $(c, 0)$, $(-c, 0)$ or $(0, c)$, $(0, -c)$, depending on whether the major axis is horizontal or vertical. An ellipse with a horizontal major axis has foci of $(c, 0)$ and $(-c, 0)$, while an ellipse with a vertical major axis has foci of $(0, c)$ and $(0, -c)$.

With the vertices represented by a, the co-vertices represented by b, and the foci represented by c, they are related by this equation:

$$c^2 = a^2 - b^2$$

If you are given the vertices and the foci or the co-vertices and the foci, you can use this equation to find the missing values (either the vertices or co-vertices).

You can find the standard form of the equation of an ellipse given a focus at $(8, 0)$ and a vertex at $(-10, 0)$. Because the focus and the vertex have an x-value other than 0 in the coordinate, the ellipse has a horizontal major axis. Therefore, the appropriate form of the equation is $\dfrac{x^2}{a^2} + \dfrac{y^2}{b^2} = 1$. The focus value is c, so $c = 8$. The vertex value is a, so $a = -10$. Substitute these values into $c^2 = a^2 - b^2$ to find b:

$$c^2 = a^2 - b^2$$
$$8^2 = -10^2 - b^2$$
$$64 = 100 - b^2$$
$$64 - 100 = 100 - 100 - b^2$$
$$-36 = -b^2$$
$$36 = b^2$$
$$\sqrt{36} = \sqrt{b^2}$$
$$6 = b$$

With $a = -10$ and $b = 6$, you can now write the equation of the ellipse:

$$\frac{x^2}{a^2} + \frac{y^2}{b^2} = 1$$

$$\frac{x^2}{(-10)^2} + \frac{y^2}{6^2} = 1$$

$$\frac{x^2}{100} + \frac{y^2}{36} = 1$$

The equation of the ellipse is $\dfrac{x^2}{100} + \dfrac{y^2}{36} = 1$.

> **HELPFUL POINT**
>
> In your future studies of mathematics, you will learn about an additional type of conic section called the *hyperbola*. A hyperbola is the intersection of a perpendicular plane with a double-napped cone, or two cones stacked on top of each other on their apex. Circles, parabolas, and ellipses can also be formed by the intersection of a plane and a single cone or a double-napped cone.

The Least You Need to Know

- A conic section is formed by the intersection of a plane with a cone. The type of conic section depends on the angle of the plane. Circles, parabolas, and ellipses are examples of conic sections.
- The equation of a circle on the coordinate plane is $(x - h)^2 + (y - k)^2 = r^2$.
- A parabola opens up or right if p is positive and opens down or left if p is negative.
- An ellipse can have a horizontal or vertical axis, which determines how you write the equation for it.

CHAPTER 19

Transformations

When is the last time you looked in the mirror or moved a pot from one gas burner to the other? You look in the mirror to see a reflection of yourself, while you move the pot from one gas burner to the next to change its location. These changes are known as *transformations*. In this chapter, you learn how to move a geometric figure to produce a new figure, as well as how to use multiple transformations.

Congruence Transformations

A transformation changes a geometric figure, resulting in a new figure. The figure being changed is known as the *pre-image,* while the resulting figure is known as the *image.* If you recall from Chapter 13, dilation is a transformation that changes the size of a geometrical figure by a reduction or an enlargement. In this case, we're going to look at changing the position of a figure without changing the size, which is known as *congruence transformation* or *isometry.* The following are three types of congruence transformations.

In This Chapter

* Translating a figure
* Reflecting a figure on any given line
* Rotating a figure about a point
* Performing combinations of two or more transformations

Translation

The first transformation we will discuss is known as a *translation*. A translation slides every point of a figure the same distance in the same direction. A translation *preserves* distance, angle measure, and *orientation*.

DEFINITION

Preserve means to not change. The **orientation** of an image is the relative position of the image.

On a coordinate plane, a translation moves a figure right or left and up or down. If the figure is translated right or left, it affects every x-coordinate of the points of a figure; if the figure is translated up or down, it affects every y-coordinate of the points of the figure. To indicate translation, you use the notation $(x, y) \rightarrow (x + h, y + v)$, which indicates point (x, y) is translated horizontally h units and vertically v units.

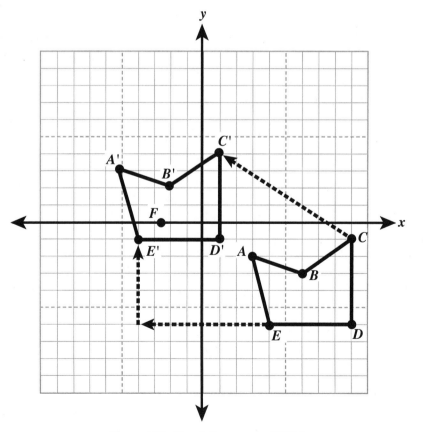

Figure 19.1: *Translating pentagon ABCDE.*

For example, in Figure 19.1, pentagon *ABCDE* has vertices *A* (3, -2), *B* (6, -3), *C* (9, -1), *D* (9, -6), and *E* (4, -6). Pentagon *ABCDE* is the pre-image. Pentagon *A'B'C'D'E'* is the image after the translation $(x, y) \rightarrow (x - 8, y + 5)$. To get this translation, you subtract 8 from the *x*-coordinates of each vertex and add 5 to each of the *y*-coordinates of each vertex to sketch the image:

$$A (3, -2) \rightarrow A' (3 - 8, -2 + 5) \rightarrow A' (-5, 3)$$

$$B (6, -3) \rightarrow B' (6 - 8, -3 + 5) \rightarrow B' (-2, 2)$$

$$C (9, -1) \rightarrow C' (9 - 8, -1 + 5) \rightarrow C' (1, 4)$$

$$D (9, -6) \rightarrow D' (9 - 8, -6 + 5) \rightarrow D' (1, -1)$$

$$E (4, -6) \rightarrow E' (4 - 8, -6 + 5) \rightarrow E' (-4, -1)$$

Another way to describe a translation is by using a vector. In the case of Figure 19.1, a vector is $\overline{CC'}$. The *initial point* or starting point of the vector is *C*, while the *terminal point* or ending point of the vector is *C'*. To write the vector component, you use the form *<h, v>*, which represents the horizontal component *h* and the vertical component *v* of the translation. For the example, you moved 8 units left and 5 units up, so the vector component is <-8, 5>.

Now that you have a handle on what a translation is, let's take a look at an example in which you are given the coordinates of an image and the translation rule used to attain the image.

Suppose the coordinates of an image are *D'* (-2, 1), *A'* (0, -1), and *N'* (2, 1) was attained by the translation $(x, y) \rightarrow (x + 5, y - 4)$ or a vector component of <5, -4>. You can determine the coordinate of the pre-image by applying the inverse of the rule to the coordinates of the image— that is, subtract 5 from the *x*-coordinate and add 4 to the *y*-coordinate:

$$D' (-2, 1) \rightarrow D (-2 - 5, 1 + 4) \rightarrow D (-7, 5)$$

$$A' (0, -1) \rightarrow A (0 - 5, -1 + 4) \rightarrow A (-5, 3)$$

$$N' (2, 1) \rightarrow N (2 - 5, 1 + 4) \rightarrow N (-3, 5)$$

Thus, the coordinates of the pre-image are *D* (-7, 5), *A* (-5, 3), and *N* (-3, 5).

Now it's time to take this information one step further. Earlier, we mentioned that a translation is a congruence transformation, meaning the shapes are the same size even though they're in different positions. Let's prove that triangle *DAN* is congruent to triangle *D'A'N'* using the SSS Postulate (see Chapter 10). To do this, sketch the coordinates of *DAN* and *D'A'N'* on a coordinate plane. Once you do, you can see that $DN = D'N' = 4$ units. Therefore, $\overline{DN}$ is congruent to $\overline{D'N'}$.

Now use the distance formula (see Chapter 4) to prove the other corresponding sides are also congruent:

$$DA = \sqrt{(-7--5)^2 + (5-3)^2} = \sqrt{8} = 2\sqrt{2}$$
$$D'A' = \sqrt{(-2-0)^2 + (1--1)^2} = \sqrt{8} = 2\sqrt{2}$$

Because $DA = D'A' = 2\sqrt{2}$ units, $\overline{DA}$ is congruent to $\overline{D'A'}$. Do the same thing for *AN* and *A'N'*:

$$AN = \sqrt{(-5--3)^2 + (3-5)^2} = \sqrt{8} = 2\sqrt{2}$$
$$A'N' = \sqrt{(0-2)^2 + (-1-1)^2} = \sqrt{8} = 2\sqrt{2}$$

Because $AN = A'N' = 2\sqrt{2}$ units, $\overline{AN}$ is congruent to $\overline{A'N'}$.

Therefore, by the SSS Postulate, triangle *DAN* is congruent to triangle *D'A'N'*.

Reflection

Another congruence transformation is a reflection. Unlike translation, reflection does not preserve orientation; instead, it only preserves angle measure and distance. A reflection "flips" every point of a figure over the same line. This line is known as the *line of reflection*. You can see some examples of reflections in Figure 19.2

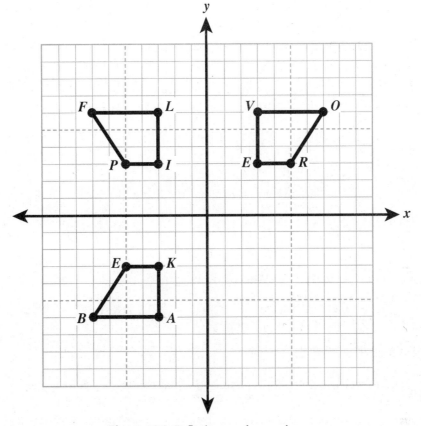

Figure 19.2: *Reflection over the x- and y-axes.*

Let's try a reflection. On a piece of notebook paper, trace the coordinate plane and trapezoid *FLIP* located in the second quadrant. Next, fold the paper over to the right or left—it does not matter which direction, as long as you align the axes of the coordinate plane. What do you notice? You should notice that trapezoid *FLIP* maps onto trapezoid *OVER* in the first quadrant. This is an example of a reflection where the line of a reflection is the *y*-axis.

Compare the coordinates of trapezoid *FLIP* to the coordinates of trapezoid *OVER:*

F (-7, 6) → O (7, 6)

L (-3, 6) → V (3, 6)

I (-3, 3) → E (3, 3)

P (-5, 3) → R (5, 3)

HELPFUL POINT

The order of the vertices in the transformation statement tells you that O is the image of F, V is the image of L, E is the image of I, and R is the image of P.

Notice how the x-coordinate of the image is -1 times the x-coordinate of the pre-image. In general, when reflecting a figure over the y-axis, the rule is $(x, y) \rightarrow (-x, y)$.

Now go back to pre-image *FLIP* and this time fold the paper up or down, once again being sure to align the axes. You should notice that trapezoid *FLIP* maps onto trapezoid *BAKE*. This is an example of a reflection over the x-axis.

Compare the coordinates of trapezoid *FLIP* to the coordinates of trapezoid *BAKE*:

$F(-7, 6) \rightarrow B(-7, -6)$

$L(-3, 6) \rightarrow A(-3, -6)$

$I(-3, 3) \rightarrow K(-3, -3)$

$P(-5, 3) \rightarrow E(-5, -3)$

In this case, you can see how the y-coordinate of the image is -1 times the y-coordinate of the pre-image. In general, when reflecting a figure over the y-axis, the rule is $(x, y) \rightarrow (x, -y)$.

But what happens when you reflect over a line other than the x- or y-axis? For example, take a look at Figure 19.3

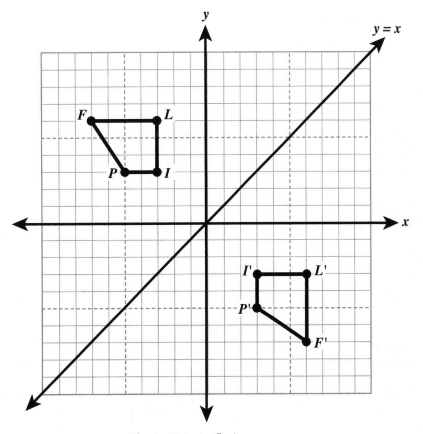

Figure 19.3: *A reflection over y = x.*

In Figure 19.3, the coordinates of trapezoid *FLIP* are reflected over the line of reflection $y = x$. To better understand this, take another piece of notebook paper and once again, trace the coordinate plane and trapezoid *FLIP*. Next, draw the line $y = x$. Fold at the line and trace the image. You should get the following:

$$F\,(-7, 6) \rightarrow F'\,(6, -7)$$

$$L\,(-3, 6) \rightarrow A\,(6, -3)$$

$$I\,(-3, 3) \rightarrow K\,(3, -3)$$

$$P\,(-5, 3) \rightarrow E\,(3, -5)$$

This time, the x-coordinate and y-coordinate switched. In general, when reflecting a figure over the line $y = x$, the rule is $(x, y) \rightarrow (y, x)$.

Here are a couple important points to understand when reflecting a figure in any given line:

- If the point you are reflecting is not on the line of reflection, the line of reflection is the perpendicular bisector of the segment connecting the point and its reflection point.

- If the point you are reflecting is on the line of reflection, the point reflects onto itself.

Take a look at Figure 19.4.

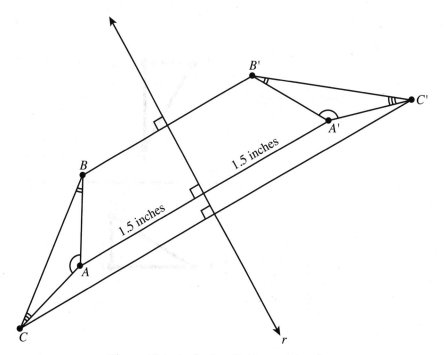

Figure 19.4: *A reflection of BAC over a given line.*

This figure shows a line segment from vertex *A* to the line of reflection *r*, so the line segment is perpendicular to the line of reflection.

The distance from the point to the line of reflection measures 1.5 inches. Therefore, you can see the line is extended perpendicular to *r* 1.5 inches on the other side of *r* to determine the location of *A'*. This then follows the same procedure to determine the location of *B'* and *C'*.

Rotation

A rotation "turns" an image about a fixed point. Like reflection, rotation preserves angle measure and distance but not orientation, in addition to orientation. As you can see in the following example, ΔABC is labeled clockwise and so is its rotation $A'B'C'$. This is due to rotations being created by two reflections—the first will reverse orientation, the second will reverse it back to the original orientation.

For example, in Figure 19.5, ΔABC located in quadrant III is the pre-image. The fixed point P, also known as the *center of rotation*, is the point about which the figure is rotating; in this case, the point is at the origin.

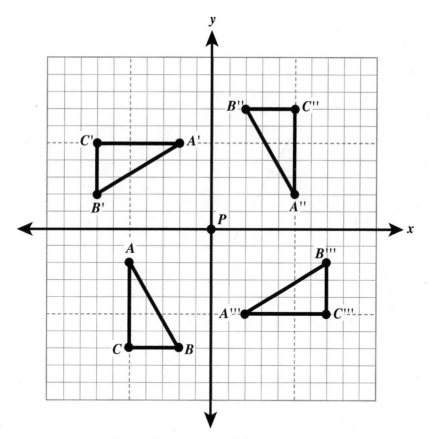

Figure 19.5: *Rotating ΔABC about the origin.*

To better understand this, take a piece of notebook paper and draw a coordinate plane with ΔABC and point P. Next, place the point of your pencil at P. Holding the pencil down so only the paper can move, turn the paper 90° clockwise in the direction of the minute hand on a clock. You can now assume ΔABC is in quadrant II. Therefore, the new coordinates are the following.

A (-5, -2) → A' (-2, 5)

B (-2, -7) → B' (-7, 2)

C (-5, -7) → C' (-7, 5)

If you turn the paper back to its original position and plot A', B', and C', you'll see it is a 90° clockwise rotation of $\triangle ABC$ about the origin.

Next, turn the paper 90° clockwise twice, from it's original position in quadrant III. Assume now that $\triangle ABC$ is in quadrant I. Therefore, the new coordinates are the following:

A (-5, -2) → A'' (5, 2)

B (-2, -7) → B'' (2, 7)

C (-5, -7) → C'' (5, 7)

If you turn the paper back to its original position and plot A'', B'', and C'', it is a 180° clockwise rotation of $\triangle ABC$ about the origin.

Finally, turn the paper 90° clockwise three times, from its original position in quadrant III. Assume now that $\triangle ABC$ is in quadrant IV. Therefore, its new coordinates are as follows:

A (-5, -2) → A''' (2, -5)

B (-2, -7) → B''' (7, -2)

C (-5, -7) → C''' (7, -5)

After turning the paper back to its original position and plotting A''', B''', and C''', you'll find it is a 270° clockwise rotation of $\triangle ABC$ about the origin.

The following are some general rules for when a point (x, y) is rotated clockwise about the origin:

Rotation of 90°: $(x, y) → (y, -x)$

Rotation of 180°: $(x, y) → (-x, -y)$

Rotation of 270°: $(x, y) → (-y, x)$

HELPFUL POINT

If you turn the paper 90° *counter-clockwise*, or in the opposite direction of the minute hand on a clock, you should notice that a 90° counter-clockwise rotation is the same as a 270° clockwise rotation.

What do you suppose happens if a figure rotates 360 °? That's right, the pre-image will map onto itself!

But what happens if your center of rotation is not the origin and not even on the coordinate plane? Take a look at Figure 19.6, which shows you how to draw a 50° counter-clockwise rotation about the point P.

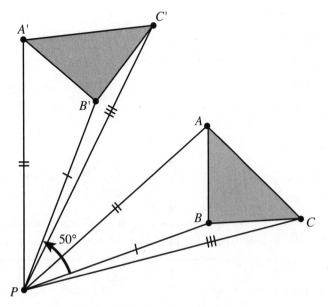

Figure 19.6: *Rotation about a point P.*

Try this by taking a protractor and a ruler and drawing a line segment from *B* to *P*. Now draw a line segment that forms a 50° angle with $\overline{BP}$ and is the length of $\overline{BP}$—that is, *BP* = *B'P'*. Repeat these steps to determine the location of *A'* and *C'*.

A rotation about a point *P* through an angle of 50° maps every vertex of $\triangle ABC$ to *A'B'C'* so that *AP* = *A'P'*, *BP* = *B'P'*, and *CP* = *C'P'*.

Composite Transformations

A composite transformation is when you combine two or more transformations. Some examples are a rotation followed by a reflection, a translation followed by a reflection, two reflections over parallel lines, and two reflections over perpendicular lines.

For example, Figure 19.7 shows a graph of the quadrilateral *WORK* with vertices *W* (3, 7), *O* (8, 5), *R* (3, 3), and *K* (5, 5).

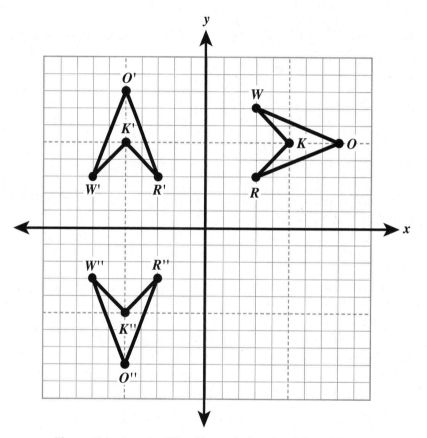

Figure 19.7: *A rotation followed by a reflection of quadrilateral WORK.*

The pre-image is then rotated 90° counter-clockwise. The coordinates for the rotated image are the following:

$$W(3, 7) \rightarrow W'(-7, 3)$$

$$O(8, 5) \rightarrow O'(-5, 8)$$

$$R(3, 3) \rightarrow R'(-3, 3)$$

$$K(5, 5) \rightarrow K'(-5, 5)$$

Finally, the new image is reflected over the *x*-axis. The coordinates for the reflection are as follows:

$$W'(-7, 3) \rightarrow W''(-7, -3)$$

$$O'(-5, 8) \rightarrow O''(-5, -8)$$

$$R'(-3, 3) \rightarrow R''(-3, -3)$$

$$K'(-5, 5) \rightarrow K''(-5, -5)$$

A rotation of *WORK* followed by a reflection of *W'O'R'K'* gave us a final image *W"O"R"K"* in the third quadrant.

HELPFUL POINT

Suppose the pre-image had first been reflected over the *x*-axis and then rotated 90° counter-clockwise. What does that give you? A final image in the first quadrant!

The following are some special composite transformations.

Glide Reflection

One type of composition transformation is a glide reflection. A glide reflection is a translation followed by a reflection. You can find an example of glide reflection in Figure 19.8.

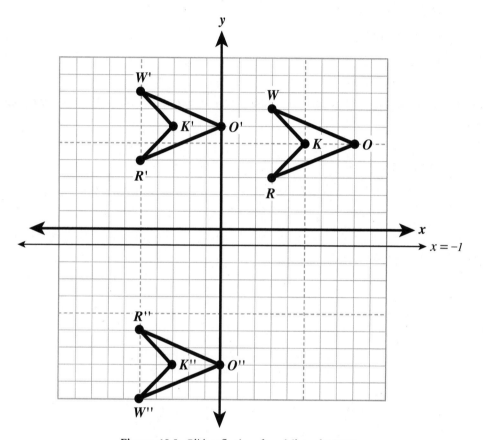

Figure 19.8: *Glide reflection of quadrilateral WORK.*

The pre-image of quadrilateral WORK has the vertices W (3, 7), O (8, 5), R (3, 3), and K (5, 5). This image is then translated <-8, 1>, which leads to the following coordinates:

W (3, 7) $\rightarrow$ W'(-5, 8)

O (8, 5) $\rightarrow$ O'(0, 6)

R (3, 3) $\rightarrow$ R'(-5, 4)

K (5, 5) $\rightarrow$ K'(-3, 6)

The translated image is then reflected over $x = -1$. The coordinates of the reflected image are as follows:

$W'(-5, 8) \rightarrow W''(-5, -10)$

$O'(0, 6) \rightarrow O''(0, -8)$

$R'(-5, 4) \rightarrow R''(-5, -6)$

$K'(-3, 6) \rightarrow K''(-3, -8)$

If you reflect the pre-image over $x = -1$, then translate it $<-8, 1>$ you will end up with the same final image.

Two Reflections

Doing two reflections is a composite transformation that can result in a translation or a rotation. Let's go over each type.

A composition of two reflections over parallel lines results in a translation. Figure 19.9 shows two reflections of $\overline{AB}$ over parallel lines, $\overleftrightarrow{m}$ and $\overleftrightarrow{n}$.

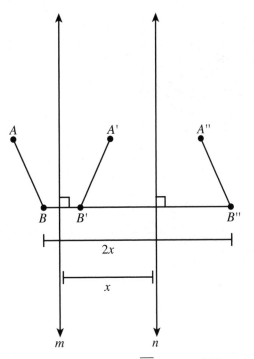

Figure 19.9: *Reflection of $\overline{AB}$ over parallel lines.*

First, $\overline{AB}$ is reflected over $\overrightarrow{m}$ using a perpendicular bisector. Remember, for example, that B' must be the same distance from $\overrightarrow{m}$ as B is from $\overrightarrow{m}$. Next, $\overline{A'B'}$ is reflected over $\overrightarrow{n}$, resulting in $\overline{A"B"}$. As you can see, the final image is a translation of the pre-image!

If x represents the distance from $\overrightarrow{m}$ to $\overrightarrow{n}$, then the equation for this final image is as follows:

$$BB" = 2x$$

Therefore, the distance from each vertex of the pre-image to the final image is twice the distance between the parallel lines.

A composition of two reflections over intersecting lines results in a rotation. For example, Figure 19.10 shows two reflections of $\overline{AB}$ over intersecting $\overrightarrow{m}$ and $\overrightarrow{n}$.

HELPFUL POINT

A composition of two reflections over parallel lines is equivalent to a single translation, while a composition of two rotations over intersecting lines is equivalent to a single rotation.

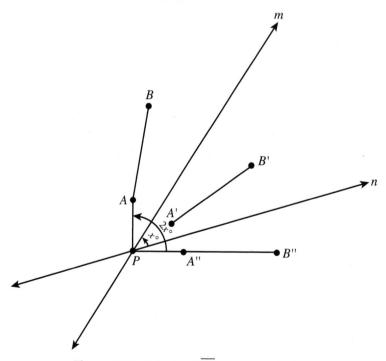

Figure 19.10: *Reflection of $\overline{AB}$ over intersecting lines.*

First, $\overline{AB}$ is reflected over $\overrightarrow{m}$ using a perpendicular bisector. Next, $\overline{A'B'}$ is reflected over $\overleftrightarrow{n}$, resulting in $\overline{A''B''}$. As you can see, the final image is a rotation of the pre-image!

If x degrees represents the measure of the acute or right angle formed by $\overrightarrow{m}$ and $\overleftrightarrow{n}$, the angle of rotation is $2x$ degrees. Therefore, the angle of rotation about point P is twice the acute or right angle formed by $\overrightarrow{m}$ and $\overleftrightarrow{n}$.

The Least You Need to Know

- Translation slides a figure while preserving its angle measure, distance, and orientation.
- With reflection, you flip a figure. This preserves angle measure and distance but changes the orientation.
- Rotation turns a figure while preserving its angle measure and distance.
- A composite reflection is combination of two or more transformations.
- Two reflections over parallel lines results in a translation, while two reflections over intersecting lines results in a rotation.

PART 8

Practice Problems

Here are 175 practice problems covering topics from each of the chapters. Although knowing a skill is important, these questions serve as applications of the skill, which is more important. In some cases, you may have to apply more than one skill or concept. This application is referred to as problem-solving, and the more you practice the better you become at test-taking. Be your own teacher! Do the problems and grade yourself. The answers are found in Appendix B. Remember, you can always reference the chapters if you are not sure what to do.

Practice Problems

Use the following figure to answer questions 1 through 5.

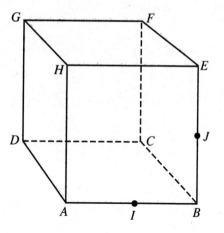

1. Which set of points is collinear?

   A. *D, G, H*

   B. *H, C, B*

   C. *E, J, B*

   D. *A, D, C, B*

2. Which set of points is coplanar?

   A. *A, I, D, G*

   B. *A, H, E, F*

   C. *B, C, D, A*

   D. *G, H, B, J*

3. Which lines are skew?

   A. $\overleftrightarrow{DG}$ and $\overleftrightarrow{CF}$

   B. $\overleftrightarrow{HE}$ and $\overleftrightarrow{EB}$

   C. $\overleftrightarrow{DG}$ and $\overleftrightarrow{GF}$

   D. $\overleftrightarrow{DG}$ and $\overleftrightarrow{HE}$

4. Name two different rays with endpoint *B*.

5. Name a plane that is parallel to plane *DAHG*.

Use the following figure to answer questions 6 and 7.

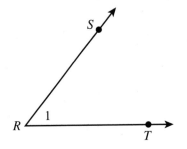

6. The measure of $\angle R$ is approximately what?

   A. 50°

   B. 130°

   C. 12°

   D. 90°

7. Which name for the angle shown is not correct?

   A. $\angle SRT$

   B. $\angle 1$

   C. $\angle R$

   D. $\angle RST$

Use the following figure to answer questions 8 and 9.

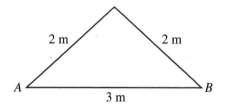

8. Classify the triangle by its sides.

9. The triangle has _____ congruent angles.

   A. 0

   B. 1

   C. 2

   D. 3

Use the following figure to answer questions 10 through 12.

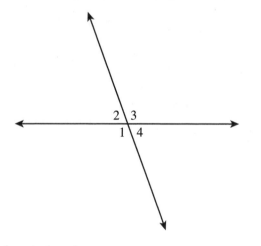

10. Name a pair of vertical angles.

11. Name a pair of supplementary angles.

12. If $m\angle 1 = 110°$, then $m\angle 3 =$ _____.

13. If two angles are complementary, then the sum of their angles is _____.

Use the following figure to answer questions 14 and 15.

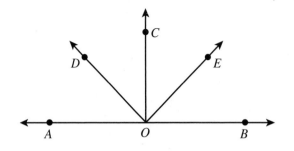

14. If $\overrightarrow{OC}$ bisects $\angle AOB$, then the measure of $\angle COB$ is _____.

15. If $\overrightarrow{OD}$ bisects $\angle AOC$, which of the following must be true?

   A. $\angle COE \cong \angle EOB$

   B. $\angle DOC \cong \angle COE$

   C. $\angle AOD$ and $\angle BOE$ are complementary

   D. $\angle AOD \cong \angle DOC$

16. In the following figure, $\angle A \cong \angle B$. Which of the following statements must be true?

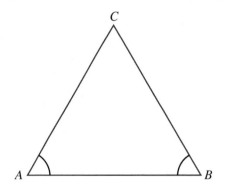

   A. $\overline{AB} \cong \overline{AC}$
   B. $\overline{AC} \cong \overline{BC}$
   C. $\overline{BC} \cong \overline{AB}$
   D. $\overline{AC} \cong \overline{BA}$

17. In an equilateral triangle, the measure of each angle is _____.

18. The lengths of two sides of a triangle are 5 and 9. Which of the following could be the length of the third side?

A. 2

B. 13

C. 15

D. 45

19. The measure of each base angle of an isosceles triangle is 15°. The measure of the vertex angle is _____.

Use the following figure to answer questions 20 through 22.

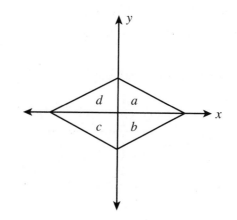

20. Which of the triangles shows how Δ*a* would look reflected about the *x*-axis?

A. *a*

B. *b*

C. *c*

D. *d*

21. Which of the triangles shows how Δ*d* would look reflected about the *y*–axis?

A. *a*

B. *b*

C. *c*

D. *d*

22. Which of the triangles shows how Δ*c* would look rotated 180° counter-clockwise about the origin?

    A. *a*

    B. *b*

    C. *c*

    D. *d*

23. The sides of a triangle are 7, 10, and 12 centimeters long. What is the classification for the triangle?

    A. Acute

    B. Obtuse

    C. Right

    D. Isosceles

24. Using the following figure, what is the intersection of $\overrightarrow{DE}$ and plane *FAC*?

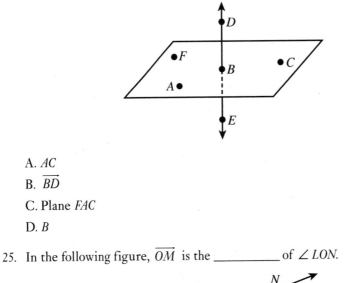

    A. *AC*

    B. $\overrightarrow{BD}$

    C. Plane *FAC*

    D. *B*

25. In the following figure, $\overrightarrow{OM}$ is the _____ of ∠*LON*.

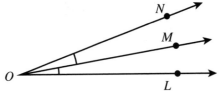

    A. Midpoint

    B. Segment bisector

    C. Angle bisector

    D. Perpendicular bisector

26. If two sides of a triangle are 3 and 4, what must the measure of the third side be for the triangle to be a right triangle?

27. In the following figure, what is $m\angle 1$?

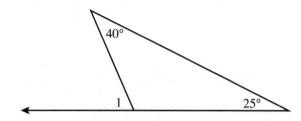

28. Which property of equality or congruence justifies the statement, "If $3(m\angle 3) = 90°$, then $m\angle 3 = 30°$"?

    A. Reflexive property of equality

    B. Distributive property

    C. Commutative property of equality

    D. Addition property of equality

    E. Multiplication property of equality

29. Which property of equality or congruence justifies the statement, "If $x = y$, then $y = x$"?

    A. Reflexive property of equality

    B. Distributive property

    C. Symmetric property of equality

    D. Addition property of equality

    E. Multiplication property of equality

30. Which property justifies the statement, "$ab - ac = a(b - c)$"?

    A. Reflexive property of equality

    B. Distributive property

    C. Commutative property of equality

    D. Addition property of equality

    E. Multiplication property of equality

Use the following figure to answer questions 31 and 32.

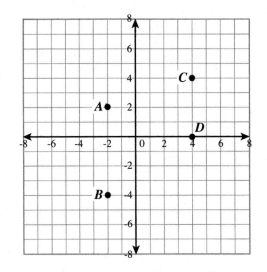

31. What is the image of B under the translation $(x, y) \rightarrow (x + 3, y - 2)$?

    A. (-1, -4)

    B. (1, -2)

    C. (-1, 4)

    D. (1, -6)

32. What is the translation from D to A?

33. In $\triangle EFG$, if $m \angle E = 35°$ and $m \angle F = 40°$, what is $m \angle G$?

34. In the following figure, which side of the triangle is the shortest?

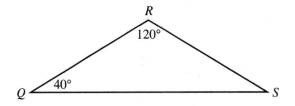

35. In the following triangle, which angle of the triangle is the largest?

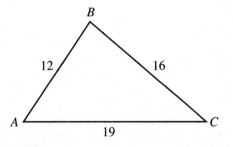

36. In the following figure, find the approximate length of *x*.

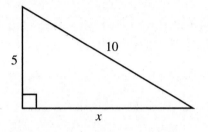

Use the following figure to answer questions 37 through 46.

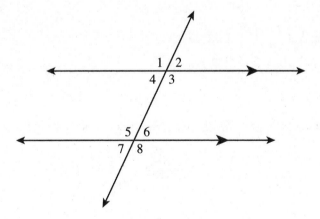

If $m \angle 4 = 100°$, find the measure of the following angles.

37. $\angle 1$

38. $\angle 2$

39. $\angle 6$

40. $\angle 8$

Determine if the following pairs of angles are vertical angles, corresponding angles, alternate interior angles, same-side interior angles, or alternate exterior angles.

41. $\angle 1$ and $\angle 3$

42. $\angle 3$ and $\angle 5$

43. $\angle 2$ and $\angle 6$

44. $\angle 4$ and $\angle 5$

45. $\angle 3$ and $\angle 8$

46. $\angle 1$ and $\angle 8$

For questions 47 through 53, determine if the triangles are congruent by SSS Postulate, SAS Postulate, ASA Postulate, AAS Postulate, or HL Theorem.

47.

48.

49.

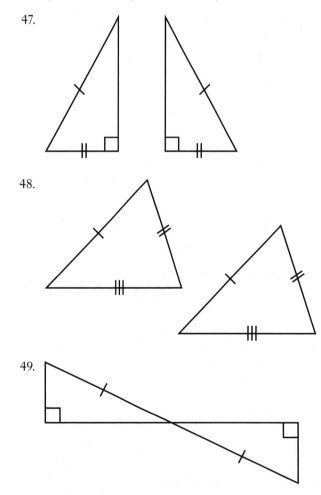

50.

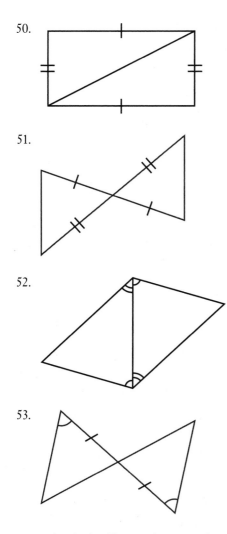

51.

52.

53.

54. What is the distance between the points (-2, 3) and (4, 3)?

55. What is the midpoint of $\overline{AB}$ with A (-1, 5) and B (6, -3)?

56. In the following figure, find the value of y if $m\angle AMC = 70°$.

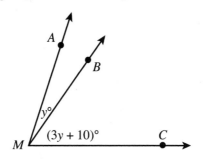

Use the following figure to answer questions 57 and 58.

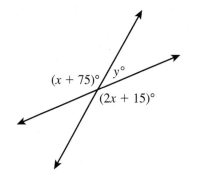

57. Find x.

58. Find y.

Use the following figure to answer questions 59 through 61.

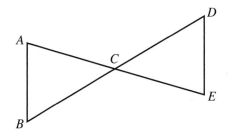

59. If $GI = 51$, find the value of y.

60. Find the length of $\overline{GH}$.

61. Find the length of $\overline{HI}$.

62. Complete the following proof.

Given: $\overline{AC} \cong \overline{CE}$ and C is the midpoint of $\overline{BD}$

Prove: $\angle A \cong \angle E$

Statements	Reasons

For questions 63 through 66, graph a polygon with vertices L (2, -2), M (4,2), N (-3,2), and P (-1, -2).

63. What is the perimeter of polygon $LMNP$?

64. What is the area of polygon $LMNP$?

65. Dilate the polygon with a scale factor of 3. What is the perimeter of the enlarged polygon?

66. What is the area of the enlarged polygon?

67. A boy knows his height is 6 feet and his shadow is 48 inches long. At the same time of day, a tree's shadow is 24 feet long. What is the height of the tree?

68. Given $\triangle ABC$ is similar to $\triangle XYR$ in the following figure, what is the measure of $\angle Y$?

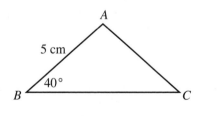

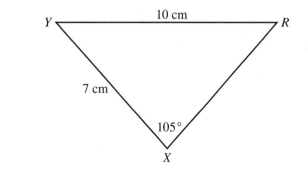

69. A pentagon has side measures $x + 5$ and $4x - 7$. If each of the measures is in terms of inches, what is the perimeter of the pentagon in feet?

    A. 4 feet

    B. 5 feet

    C. 3 feet

    D. 3.75 feet

70. The midpoint of $\overline{XZ}$ is $Y(4, -2)$. If one endpoint is $X(0, -2)$, what is the length of $\overline{XZ}$?

    A. 4 units

    B. 8 units

    C. 2 units

    D. 10 units

71. A pair of same-side interior angles measures $4x + 6$ and $2x + 6$. What is a possible angle measure for one of these angles?

    A. $12°$

    B. $28°$

    C. $62°$

    D. $180°$

72. In the following diagram, m is parallel to n. Which pair of angle measures is equal?

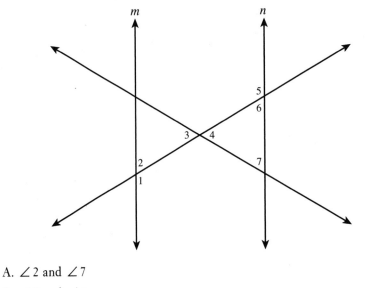

    A. $\angle 2$ and $\angle 7$

    B. $\angle 2$ and $\angle 6$

    C. $\angle 3$ and $\angle 7$

    D. $\angle 1$ and $\angle 3$

73. Given the true conditional, "If it rains, then you will need an umbrella," which of the following related conditionals must also be true?

    A. Converse

    B. Inverse

    C. Contrapositive

    D. Biconditional

74. In the following figure, trapezoid *ABCD* is similar to trapezoid *EFCG*. Find the value of *x*.

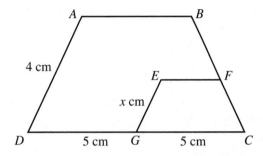

75. Supplementary angles are two angles whose measures add up to what number of degrees?

    A. 180°

    B. 90°

    C. 360°

    D. 270°

76. Perpendicular lines are two lines that intersect at what degree angle?

    A. 180°

    B. 90°

    C. 360°

    D. 270°

77. Which quadrilateral has exactly one pair of opposite sides that are congruent?

    A. Kite

    B. Trapezoid

    C. Isosceles trapezoid

    D. Rectangle

78. What is the measure of each angle in a regular decagon in degrees?

    A. 180°

    B. 144°

    C. 360°

    D. 145°

79. How many diagonals can you draw from one vertex of a hexagon?

    A. 6

    B. 5

    C. 4

    D. 3

80. Polygons are congruent if their _____ angles and sides are congruent.

    A. Supplementary

    B. Corresponding

    C. Vertical

    D. Alternate interior

81. A cube has a total surface area of 384 square units. What is the length of one edge?

    A. 128 units

    B. 96 units

    D. 8 units

    E. 64 units

82. The area of a rectangle is 4,800 cm² and one dimension is 60 cm. Find the length of the diagonal of the rectangle.

83. In the following figure, what is the measure of ∠ Z?

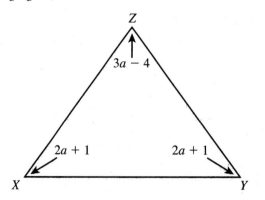

For questions 84 through 89, find the measure of the angle given $m\angle 1 = (5x + 15)°$, $m\angle 2 = (28x)°$, $m\angle 3 = (45y)°$, $m\angle 4 = (7n + 39)°$, $m\angle 5 = (11n - 13)°$, $m\angle 6 = (110 - 10y)°$.

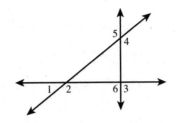

84. $m\angle 1$

85. $m\angle 2$

86. $m\angle 3$

87. $m\angle 4$

88. $m\angle 5$

89. $m\angle 6$

90. The length of $AB = 10$ in. The length of $A'B' = 5$ in. What is the scale factor?

91. If trapezoid $ABCD$ is similar to trapezoid $EFGH$, which of the following statements must be true?
    A. $\overline{AD} \cong \overline{FG}$
    B. $\angle C \cong \angle H$
    C. $\angle D \cong \angle H$
    D. $\overline{BC} \cong \overline{EH}$

For questions 92 through 97, use the following figure.

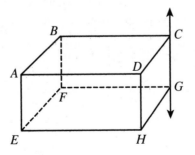

92. Name two lines parallel to $\overleftrightarrow{DH}$.

93. Name two lines skew to $\overrightarrow{EF}$.

94. Name the plane parallel to CDG.

95. Name a pair of perpendicular planes.

96. Write a complete two-column proof.
    Given: $\overline{BE} \perp \overline{AD}$; $\angle CDE$ is a right angle
    Prove: $\angle 4 \cong \angle 3$

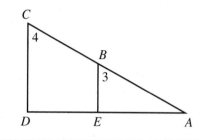

Statements	Reasons

For questions 97 through 100, determine if the two triangles are congruent. If they are, name the triangle congruence theorem or postulate you would use.

97.

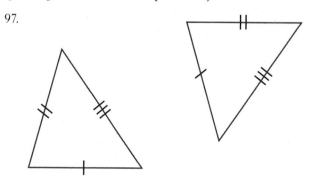

98.

99.

100.

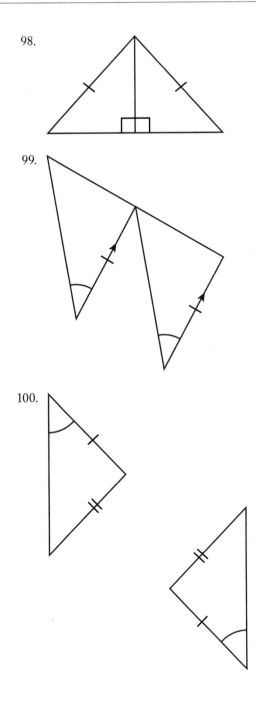

For questions 101 through 104: Q is the midpoint of $\overline{RT}$ and $RT = 26$. Fill in each blank with the correct number.

101. $RQ =$ _____

102. $TQ =$ _____

103. $TQ =$ _____ $\times RT$

104. $\dfrac{RQ}{RT} =$ _____

For questions 105 through 107, say whether each statement is true or false. If it is false, provide a counterexample.

105. $x^2 > 9$ implies $x > 3$.

106. If a four-sided figure has four congruent sides, then it has four right angles.

107. If the height of a cylinder is 7 feet and the radius of the base is 3 feet, then the volume of the cylinder is approximately 198 feet^3.

For questions 108 through 110, say whether the shape is convex or concave.

108.

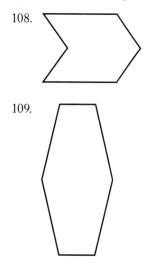

109.

110.

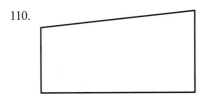

For questions 111 through 114, $GK = 24$, $HJ = 10$, and $GH = HI = IJ$. Find each length.

111. HI

112. JK

113. IG

114. IK

For question 115, find the value of x.

115.

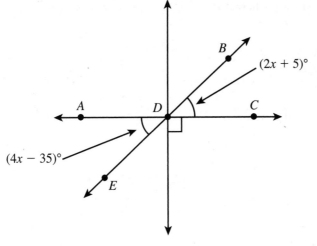

116. $\angle A$ and $\angle B$ are a linear pair where $m\angle A = \dfrac{1}{2}x + 20$ and $m\angle B = \dfrac{5}{6}x$

117. A supplement of an angle is six times as large as the complement of the angle. Find the measure of the angle, its supplement, and its complement.

118. Write a complete two-column proof.

Given: $DW = ON$

Prove: $DO = WN$

Statements	Reasons

119. Write a complete two-column proof.

Given: $m\angle 4 + m\angle 6 = 180$; $\angle 4$ *and* $\angle 5$ are a linear pair

Prove: $m\angle 5 = m\angle 6$

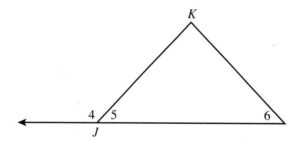

Statements	Reasons

120. On a coordinate plane, plot the ordered pairs A (-5, 1), B (-4, 5), and C (-2, 3). Connect the ordered pairs. Reflect the image over $y = x$. Translate the new image 2 units to the right and 1 unit down. Now rotate the image 90° clockwise. What are the coordinates of $A'''B'''C'''$?

121. A quadrilateral has angle measures of $2x + 9$, $11x - 17$, $8x - 3$, and $4x - 4$. What is the value of x?

For questions 122 through 126, match the term with the corresponding properties choices, A through E. There may be more than one property per polygon.

    A. Two pairs of opposite sides are $\cong$.

    B. Only one pair of opposite sides is parallel.

    C. The sum of the interior angles is 360°.

    D. The diagonals are $\perp$.

    E. The diagonals bisect each other.

122. Square

123. Trapezoid

124. Rhombus

125. Quadrilateral

126. Kite

127. $\overline{MN}$ is the midsegment of the following trapezoid. Solve for x.

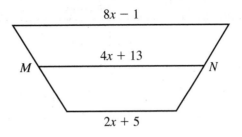